Evaluation of Streambed Scour at Bridges over Tidal Waterways in Alaska

By Jeffrey S. Conaway and Paul V. Schauer

Prepared in cooperation with the Alaska Department of Transportation and Public Facilities

Scientific Investigations Report 2012–5245

U.S. Department of the Interior
U.S. Geological Survey

U.S. Department of the Interior
KEN SALAZAR, Secretary

U.S. Geological Survey
Marcia K. McNutt, Director

U.S. Geological Survey, Reston, Virginia: 2012

For more information on the USGS—the Federal source for science about the Earth, its natural and living resources, natural hazards, and the environment, visit http://www.usgs.gov or call 1–888–ASK–USGS.
For an overview of USGS information products, including maps, imagery, and publications, visit http://www.usgs.gov/pubprod

To order this and other USGS information products, visit http://store.usgs.gov

Suggested citation:
Conaway, J.S., and Schauer, P.V., 2012, Evaluation of streambed scour at bridges over tidal waterways in Alaska: U.S. Geological Survey Scientific Investigations Report 2012-5245, 38 p.

Contents

Figures

Tables

Tables—Continued

Conversion Factors, Datums, and Abbreviations and Acronyms

Conversion Factors

Multiply	By	To obtain
Length		
foot (ft)	0.3048	meter (m)
foot (ft)	304.8	millimeter (m)
mile (mi)	1.609	kilometer (km)
Volume		
cubic foot (ft^3)	0.02832	cubic meter (m^3)
Flow rate		
cubic foot per second (ft^3/s)	0.02832	cubic meter per second (m^3/s)
Pressure		
pound per square foot (lb/ft^2)	0.04788	kilopascal (kPa)
Density		
slug per cubic foot (lb/ft^3)	515.378	kilogram per cubic meter (kg/m^3)

Datums

Vertical coordinate information is referenced to the local bridge datum used by the Alaska Department of Transportation and Public Facilities. Some tidal elevations were not translated to the local bridge datum and are referenced to Mean Sea Level (MSL).

Horizontal coordinate information is referenced to the World Geodetic Standard of 1984 (WGS 84).

Abbreviations and Acronyms

ADCP	acoustic Doppler current profiler
ADOT&PF	Alaska Department of Transportation and Public Facilities
AEPWL	Annual exceedance probability water level
BN	bridge number
HEC	Hydraulic Engineering Circular
HEC-RAS	Hydraulic Engineering Center River Analysis System
USGS	U.S. Geological Survey

Evaluation of Streambed Scour at Bridges over Tidal Waterways in Alaska

By Jeffrey S. Conaway and Paul V. Schauer

Abstract

The potential for streambed scour was evaluated at 41 bridges that cross tidal waterways in Alaska. These bridges are subject to several coastal and riverine processes that have the potential, individually or in combination, to induce streambed scour or to damage the structure or adjacent channel. The proximity of a bridge to the ocean and water-surface elevation and velocity data collected over a tidal cycle were criteria used to identify the flow regime at each bridge, whether tidal, riverine, or mixed, that had the greatest potential to induce streambed scour.

Water-surface elevations measured through at least one tide cycle at 32 bridges were correlated to water levels at the nearest tide station. Asymmetry of the tidal portion of the hydrograph during the outgoing tide at 12 bridges indicated that riverine flows were stored upstream of the bridge during the tidal exchange. This scenario results in greater discharges and velocities during the outgoing tide compared to those on the incoming tide. Velocity data were collected during outgoing tides at 10 bridges that experienced complete flow reversals, and measured velocities during the outgoing tide exceeded the critical velocity required to initiate sediment transport at three sites.

The primary risk for streambed scour at most of the sites considered in this study is from riverine flows rather than tidal fluctuations. A scour evaluation for riverine flow was completed at 35 bridges. Scour from riverine flow was not the primary risk for six tidally-controlled bridges and therefore not evaluated at those sites. Field data including channel cross sections, a discharge measurement, and a water-surface slope were collected at the 35 bridges. Channel instability was identified at 14 bridges where measurable scour and or fill were noted in repeated surveys of channel cross sections at the bridge. Water-surface profiles for the 1-percent annual exceedance probability discharge were calculated by using the Hydrologic Engineering Center's River Analysis System model, and scour depths were calculated using methods recommended by the Federal Highway Administration. Computed contraction-scour depths were greater than 2.0 feet at five bridges and computed pier-scour depths were 4.0 feet or greater at 15 bridges.

The potential for streambed scour by both coastal and riverine processes at the bridges considered in this study were evaluated, ranked, and summed to determine a cumulative risk factor for each bridge. Possible factors that could mitigate the scour risks were investigated at 22 bridges that had high individual or cumulative rankings. Mitigating factors such as piers founded in bedrock, deep pier foundations relative to scour depths, and lack of observed scour during field measurements were documented for 13 sites, but additional study and monitoring is needed to better quantify the streambed scour potential for nine sites. Three bridges prone to being affected by storm surges will require more data collection and possibly complex hydrodynamic modeling to accurately quantify the streambed scour potential. Continuous monitoring of water-surface and streambed elevation at one or more piers is needed for two bridges to better understand the tidal and riverine influences on streambed scour.

Introduction

Alaska, with its nearly 34,000 mi of coast, has more than twice the shoreline of the continental United States and tidal ranges that are the second largest in North America. The state's limited coastal infrastructure is subject to the effects of tidal fluctuations, storm surges, littoral drift, wave action, and ice. In addition to these processes, bridges over tidal waterways can be subject to riverine processes that can induce streambed scour, which is the major cause of bridge failure in the United States (Murillo, 1987).

After several tragic bridge failures in the late 1980s, the Federal Highway Administration recommended that every bridge over a waterway be evaluated for susceptibility to streambed scour (U.S. Department of Transportation, 1988). In response to that recommendation, the U.S. Geological Survey (USGS), in cooperation with the Alaska Department of Transportation and Public Facilities (ADOT&PF) began studying the susceptibility of Alaskan bridges to streambed scour in 1994. A multi-phase approach was applied to bridges selected by ADOT&PF as potentially subject to scour. Heinrichs and others (2001) documented procedures and results from the initial phase of this project at 325 bridges.

Conaway (2004) selected 54 bridges from the initial study and performed a more intensive analysis of scour susceptibility. Both studies followed the guidance outlined in the Hydraulic Engineering Circular (HEC)-18 (Richardson and others, 1993) and included estimates for contraction and pier scour at each bridge computed from one-dimensional hydraulic models of the 1- and 0.2-percent annual exceedance probability (AEP) flows corresponding to the 100- and 500-year recurrence-interval peak flows. The complexity of tidal hydraulics and limited federal guidance on analysis procedures exempted bridges over tidal waterways from these initial scour analyses. The fourth edition of HEC-18 (Richardson and Davis, 2001) provided the first guidance for evaluating bridges over tidal waterways and HEC-25 (Zevenbergen and others, 2004; Douglass and Krolak, 2008) expanded upon this work.

Bridges over tidal waterways are subject to several coastal processes that have the potential to induce streambed scour or damage the structure or roadway approaches to the structure. These processes include scour during tidal exchanges or storm surges, scour and erosion due to hydrodynamic loading from waves, and erosion or deposition from long-shore drift. Most bridges in this study cross rivers and creeks rather than estuaries or tidal inlets; therefore, the local topography limits the effects of these coastal processes.

Bridges over tidal waterways also can be subject to riverine processes that induce streambed scour at a bridge. Streambed scour at bridges results from the complex hydraulic conditions created either by the contraction of flow through the bridge or by the interaction of flow with bridge piers or abutments that results in the hydraulic erosion of the streambed or stream banks. Streambed scour is commonly separated into three processes: long-term degradation of the channel, contraction scour, and local scour. Long-term degradation of the channel accounts for the natural channel degradation that would occur regardless of the bridge. Contraction scour results from the decrease in channel width caused by the presence of the bridge and the attendant increase in flow velocity and the potential for sediment transport in this area. Local scour at piers results from horseshoe and wake vortices that form at the upstream, downstream, and sides of piers and from flow acceleration at the pier obstruction. There is a vast amount of literature on the subject of streambed scour at bridges; Richardson and Lagasse (1999) have edited a compendium on the subject that provides a thorough overview of historical and recent research.

Tides and storm surges can intensify scour conditions at a bridge by temporarily increasing the volume of water upstream of the bridge from both backwater effects and tidal flooding. The bridge can potentially act as a constriction to the outgoing flow, thereby creating a head difference between the area upstream of the bridge and the ocean. The potential for storage upstream of the bridge and the degree to which the bridge opening limits the return of stored volume back to the ocean are important factors in the assessment of scour.

In 2009, the USGS, in cooperation with ADOT&PF, began an investigation to evaluate streambed scour potential at 41 bridges over tidal waterways in Alaska (fig. 1, table 1). The objectives of this investigation were to (1) determine the degree of tidal influence at each bridge; (2) identify the flow regime (tidal or riverine) that presents the greatest risk to the bridge and determine the appropriate method to evaluate this risk; (3) measure riverine discharge, velocity, and water-surface elevations during a tidal cycle, and channel cross sections at selected sites; (4) investigate long-term degradation from repeated channel surveys; and (5) estimate the scour potential for the 1-percent AEP riverine flow using the current scour evaluation techniques in HEC-18 (Richardson and Davis, 2001) and HEC-25 (Douglas and Krolak, 2008).

There are varying degrees of complexity associated with the hydrologic and hydraulic processes at bridges over tidal-affected waterways. Relatively simple one-dimensional steady flow models are sufficient for evaluating hydraulic variables at some sites, but at other sites complex data intensive, multi-dimensional hydrodynamic models may be required to determine these same variables. The approach taken in this study was to determine the level of complexity required for an individual site and then utilize the simplest evaluation method available to accurately evaluate the potential for streambed scour. Several sites would require multi-dimensional hydrodynamic models to fully evaluate the hydrologic and hydraulic processes, but if these processes do not have the potential to scour the streambed at the bridge, then a complex model is unwarranted. Sites that cannot be evaluated fully with the procedures outlined in this study are recommended for more intensive study.

This report describes (1) the techniques developed to evaluate streambed scour at 41 bridges over tidal waterways in Alaska; (2) the results of these analyses; and (3) potential mitigating factors at high risk sites and recommendations for further analysis. Contraction and pier scour were evaluated at each bridge using variables computed from one-dimensional steady state step-backwater models. Abutment scour was not evaluated because all of the bridges in this study have abutments that are either founded in bedrock or armored by riprap or sheet piling. Hydraulic models were constructed from existing data and field data collected for this study. The degree of tidal influence at a bridge was quantified using water-stage data collected over one or more tidal cycles. Velocity data were collected during the falling tide at bridges that experienced flow reversals during tidal exchanges.

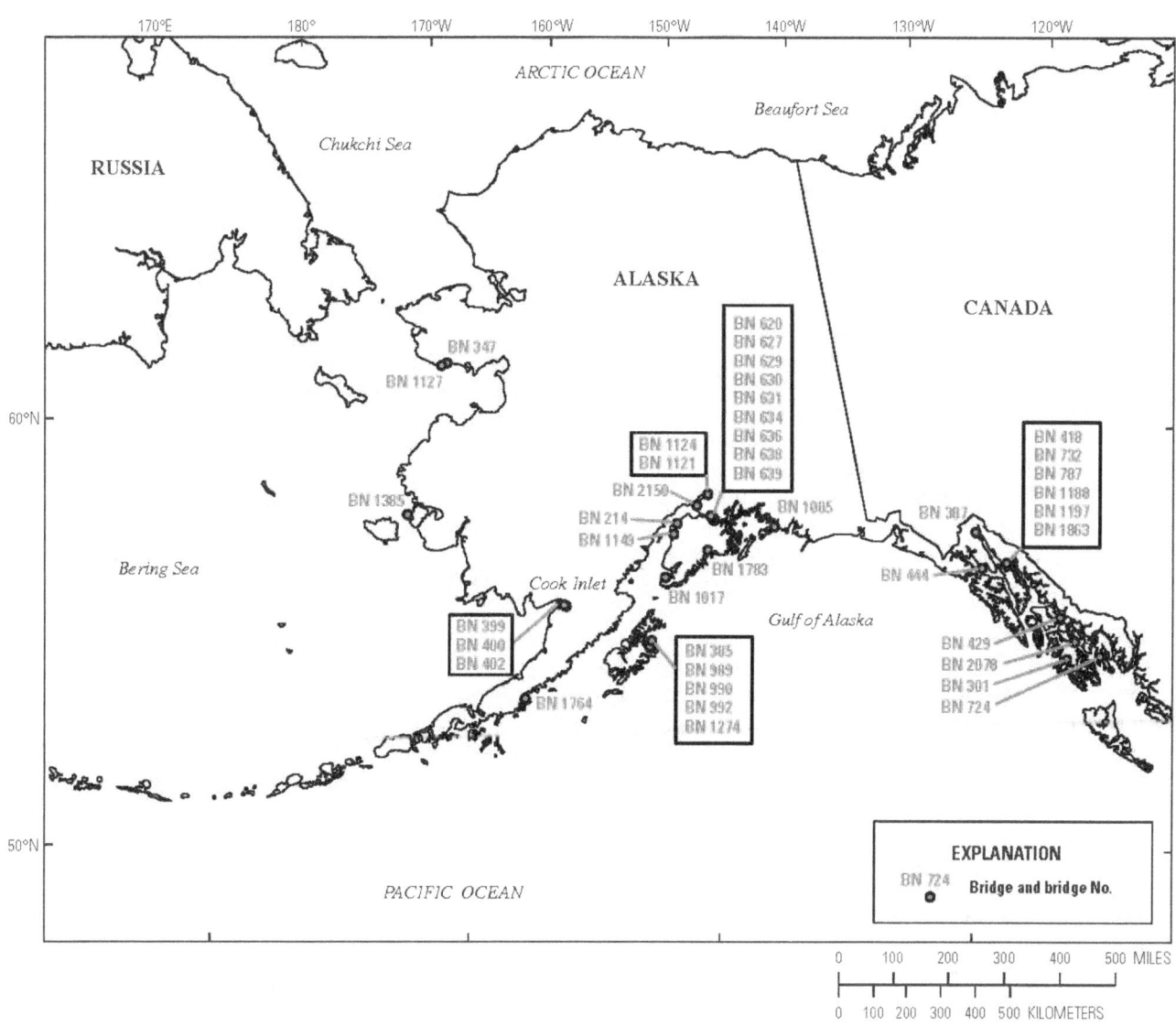

Figure 1.   Location of bridges over tidal waterways in Alaska selected for analysis of susceptibility to streambed scour.

Table 1.   Bridges over tidal waterways in Alaska selected for streambed-scour analysis and characteristics of riverine and tidal conditions at the bridges.

[Tide elevations are the maximum and minimum since 1901. **Bold** values indicate tides referenced to mean sea level. **Abbreviations:** WGS, World Geodetic Standard of 1984; ft³/s, cubic foot per second; ft, foot; mi, mile]

Bridge No.	Bridge name	Latitude (WGS 84)	Longitude (WGS 84)	Magnitude of 1-percent annual exceedance probability discharge (ft³/s)	Tide elevation (ft) Minimum	Tide elevation (ft) Maximum	Tidal classification	Nearest tidal station	Station type	Distance to tidal station (mi)
214	Swanson River	60°47'57.68"N	151° 047.47"W	3,040	-14.8	17	Affected	East Foreland	Subordinate	14.7
301	Klawock River	55°32'54.95"N	133°549.12"W	Only tidal	**-3.4**	**12.9**	Controlled	Craig	Subordinate	4.9
347	Bonanza Creek	64°32'13.59"N	164°29'25.72"W	Only tidal	-0.5	1.7	Controlled	Nome	Reference	27.1
385	Salt Creek	57°39'0.62"N	152°31'34.86"W	578	77.9	92.2	Controlled	Women's Bay	Reference	2.6
387	Chilkoot River	59°19'27.81"N	135°33'27.53"W	12,200	-10.0	16.4	Affected	Haines Inlet	Subordinate	7.7
399	King Salmon Creek	58°41'45.80"N	156°41'51.20"W	3,120	-0.6	9	Affected	Omakstalia Point	Subordinate	2.0
400	Leader Creek	58°44'49.18"N	156°56'37.82"W	100	-1.8	21.9	Affected	Morakas Point	Subordinate	1.0
402	Pauls Creek	58°43'28.81"N	156°46'50.21"W	1,700	-0.6	9	Affected	Omakstalia Point	Subordinate	2.0
418	Sheep Creek	58°15'37.45"N	134°19'31.91"W	1,140	-7.4	18.2	Influenced	JUNEAU	Reference	4.3
429	Blind River	56°36'54.92"N	132°49'12.64"W	2,710	-4.5	20.6	Influenced	Anchor Point	Subordinate	4.5
444	Salmon River	58°24'45.47"N	135°44'17.31"W	10,100	**-4.9**	**19.1**	Affected	Bartlett Cove	Subordinate	5.8
620	Ingram Creek	60°50'47.90"N	149° 336.56"W	5,300	-23.9	21	Influenced	Sunrise	Subordinate	13.1
627	Placer River Overflow	60°49'11.49"N	149° 09.01"W	10,200	-23.3	21.6	Affected	Sunrise	Subordinate	16.0
629	Placer River Main Crossing	60°49'2.15"N	148°59'20.10"W	10,200	-23.8	21.1	Affected	Sunrise	Subordinate	16.0
630	Portage Creek 1	60°49'15.48"N	148°58'38.39"W	7,160	-22.8	22.1	Influenced	Sunrise	Subordinate	16.0
631	Portage Creek 2	60°49'36.62"N	148°5845.81"W	7,160	-22.7	22.2	Influenced	Sunrise	Subordinate	16.0
634	Twentymile River	60°5041.46"N	148°59'22.63"W	18,420	-21.7	23.2	Affected	Sunrise	Subordinate	15.4
636	Peterson Creek	60°534.56"N	149° 255.07"W	547	-22.4	22.5	Influenced	Sunrise	Subordinate	12.9
638	Virgin Creek	60°55'58.58"N	149° 923.69"W	563	-26.2	18.7	Affected	Sunrise	Subordinate	9.5
639	Glacier Creek	60°5619.12"N	149°107.26"W	4,920	-25.3	19.5	Influenced	Sunrise	Subordinate	9.3
724	Ketchikan Creek	55°20'37.85"N	131°38'26.75"W	3,180	-3.5	20.9	Controlled	Ketchikan	Reference	0.8
732	Gold Creek	58°17'56.67"N	134°25'11.50"W	3,250	-8.0	17.9	Controlled	JUNEAU	Reference	0.1
787	Salmon Creek Twin Lakes Drive	58°19'52.17"N	134°38'19.26"W	3,290	-4.9	20.7	Influenced	JUNEAU	Reference	2.9
989	Sargent Creek	57°42'33.09"N	152°34'6.28"W	4,230	-2.8	11.5	Influenced	Women's Bay	Reference	2.7
990	Russian River	57°42'22.77"N	152°34'16.76"W	4,730	-2.9	11.4	Influenced	Women's Bay	Reference	2.7
992	Salonie River	57°41'46.76"N	152°33'37.91"W	5,090	-2.8	11.5	Influenced	Women's Bay	Reference	2.7
1017	Seldovia Slough	59°26'13.45"N	151°42'31.00"W	200	-8.8	20.5	Controlled	Seldovia	Reference	0.9
1085	Hartney Bay	60°307.18"N	145°5148.04"W	2,810	-9.2	10.9	Controlled	Gravel Point	Subordinate	4.3
1121	Knik River NB	61°28'55.87"N	149°15'11.65"W	120,000	7.1	21.4	Influenced	Anchorage	Reference	26.9
1124	Matanuska River NB	61°30'15.42"N	149°15'2.07"W	47,000	7.1	21.4	Influenced	Anchorage	Reference	27.9
1127	Safety Sound	64°28'19.73"N	164°44'47.07"W	Predominantly tidal	-0.5	1.7	Controlled	Nome	Reference	19.5
1149	Kenai River at Kenai Bridge	60°31'36.06"N	151°1232.34"W	42,300	-10.0	21.3	Affected	Kenai City Pier	Subordinate	1.8
1188	Salmon Creek at Egan Drive	58°19'49.39"N	134°28'26.33"W	3,290	-4.9	20.7	Influenced	JUNEAU	Reference	2.9
1197	Lemon Creek NB	58°212.73"N	134°30'28.64"W	6,940	-8.5	16.9	Affected	Fritz Cove	Subordinate	4.1
1274	Monashka Creek	57°50'28.60"N	152°2645.11"W	2,210	-7.5	6.77	Influenced	Kodiak	Subordinate	4.1
1385	Tununak River	60°34'40.70"N	165°167.66"W	1,220	**-1.6**	**5.3**	Controlled	Tachikuga Bay	Subordinate	74.9
1764	Indian Creek	56°17'54.16"N	158°24'39.88"W	1,390	**-2.5**	**11.7**	Influenced	Anchorage Bay	Subordinate	0.4
1783	Spruce Creek	60° 426.04"N	149°2643.61"W	3,930	**-3.5**	**13.8**	Influenced	Seward	Reference	3.0
1863	Lemon Creek SB	58°212.12"N	134°30'29.82"W	6,940	-8.5	16.9	Affected	Fritz Cove	Subordinate	4.1
2078	Deer Creek	55°54'8.24"N	132°306.72"W	810	**-4.6**	**20.2**	Affected	Karta Bay	Subordinate	8.6
2150	Ship Creek	61°13'25.41"N	149°53'33.78"W	1,900	-22.7	17.8	Controlled	Anchorage	Reference	0.7

Study Approach

Variable coastal and riverine conditions at the study sites required that a flexible approach be taken to evaluate the scour potential for bridges in the intertidal zone. This study followed the general guidelines of HEC-18 (Richardson and Davis, 2001) and applied one or more of the following levels of approach to:

1. Determine the magnitude of tides and storm surges, the stability of the crossing, and determine if the hydraulic analysis should include tidal flows, riverine flows, or both.

2. Determine hydraulic variables needed to evaluate contraction and pier scour, typically computed from a one-dimensional steady or unsteady step-backwater model.

3. Determine hydraulic variables using complex multi-dimensional hydrodynamic modeling.

All of the bridges in this study were analyzed through the second level, which corresponds to a HEC-18 Level 2 analysis. For sites where the third level of approach corresponding to a HEC-18 Level 3 analysis seemed appropriate, velocity data were collected during one tidal cycle to determine if the velocities during the outgoing tide were high enough to induce scour at the bridge sites.

Tidal Analysis

The degree to which astronomical tides affect the hydraulic conditions at a bridge is related to the magnitude of the tides, proximity of the bridge to the coast, bridge opening geometry and the area upstream where incoming tidal flow can be detained or stored (upstream storage area), and streambed slope. In Alaska, astronomical tides are semidiurnal, with two high tides and two low tides for every lunar day (24.8 hours). The two daily high tides are of similar magnitude at most sites in Alaska, but some areas experience mixed tides, whereby one high tide is significantly higher than the other. Several bridges in this study are in or near upper Cook Inlet, which has the second largest tidal range in North America.

The extent of tidal influence on flow through each bridge was determined from a variety of sources including tidal-station data, water-surface elevations measured through tide cycles, and aerial photography. These data were then used to group all of the sites into three categories (1) tidally controlled, (2) tidally affected, and (3) tidally influenced (fig. 2). Tidally-controlled bridges have a full flow reversal at every tide cycle. Tidally-affected bridges have reverse flow during some tide cycles, but tidal action is not the dominant flow condition. Tidally-influenced bridges are dominated by riverine flow and experience backwater at the bridge, but no flow reversal.

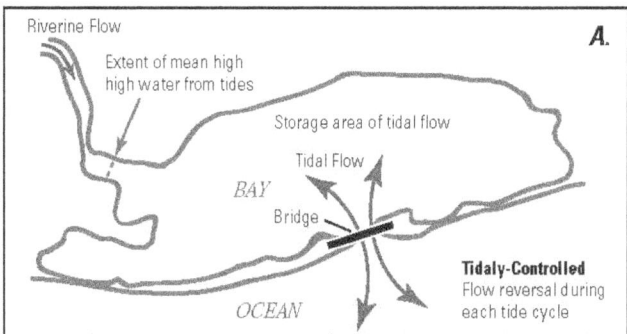

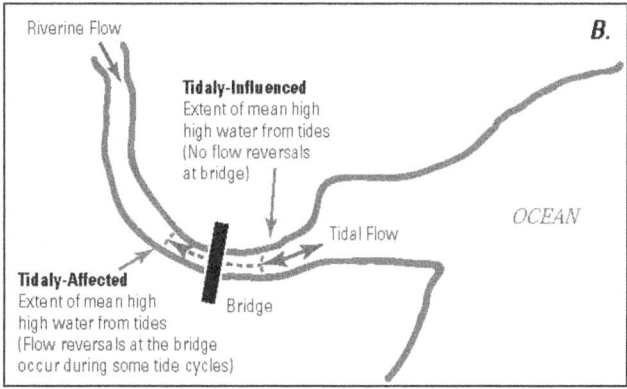

Figure 2. Classification of tidally controlled, affected, and influenced bridges.

The highest and lowest predicted tides since 1901 were determined for each bridge from data at the nearest tidal station (table 1). Tidal stations are either reference stations operated by the National Oceanic and Atmospheric Administration or subordinate stations at which tide and current predictions are calculated based on observations at the reference stations. The magnitude of tidal fluctuations at the tidally-controlled bridges are usually similar to those at the tidal stations, and the hydrograph at the bridge matches that at the tide stations for the incoming and the outgoing tides (fig. 3). The magnitude is less at the tidally-affected and tidally-influenced bridges, and the relationship between the water-surface elevations at bridges and tidal stations are dependent on several factors. Tidally-affected and tidally-influenced bridges can have a slightly different hydrograph than the tidal station on the outgoing tide, due to a combination of the riverine inflow to the bridge being stored and the outgoing tide being restricted by the bridge opening (fig. 4). During the incoming tide, however, those influences are minimized and the shape of the hydrograph at the bridge is virtually the same as that at the tide station.

Pressure transducers were installed at 33 bridges for at least one tidal cycle to develop a relation between water-surface elevations at the bridge and at the nearest tide station during an incoming tide. Transducers were not installed

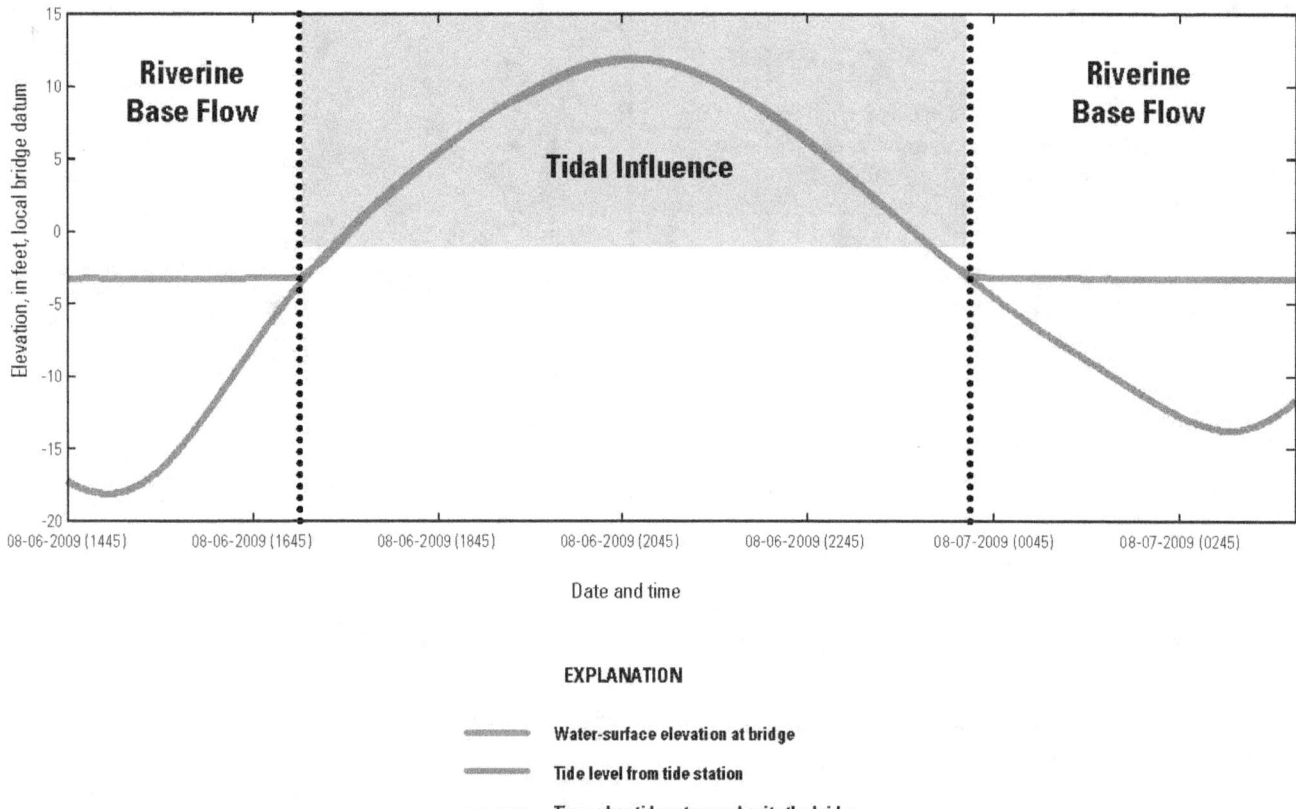

Figure 3. Water-surface elevations from the Anchorage tidal station and at the tidally-controlled bridge number 2150 over Ship Creek, Alaska. Tidal station elevations have been adjusted to the bridge datum.

at three bridges because of logistical constraints or at an additional five bridges because the affect of tides on the riverine flow was negligible. The water-surface elevations at tidally-affected and tidally-influenced bridges may not display the full signature of the incoming tide due to the elevation of the bridge above mean sea level. A minimal or no tidal signature would be expected in the water-surface elevation at these bridges for the low-tide part of the tidal cycle. The first step in the correlation between the nearest tide station and the bridge was to determine the time difference between the high-tide elevations at the tide station and at the bridge. Because the tide stations typically are not near the bridge, and the river is affected by backwater caused by the incoming tide, there is a lag between the sites. The time lag was considered to be the difference between the time of the maximum water surface at the bridge and the time of the high tide at the tide station. Tide elevations from the tide station were converted into the local bridge elevation datum (bridge datum). Elevations of the maximum high and minimum low tides (since 1901), mean higher high water (MHHW), and mean lower low

water (MLLW) were plotted on cross sections at each of the 33 bridges to determine the tidal elevations relative to the bridge structure (appendix A). The MHHW is the average of the higher high water height of each tidal day observed over a period specified by the National Ocean Service. In Alaska, tidal datums are computed on a 5-year epoch. The MLLW is the average of the lower low water height of each tidal day for the same time period. Plots of tidal elevations at the bridges do not reflect the cumulative water-surface elevations that would result by adding the tidal-elevation component to the riverine-elevation component shown individually in the plots (tide plus riverine). These components are not strictly additive because of varying channel geometries, valley slopes, discharges, and backwater effects. At several locations, the mean low-low water elevation is below the surveyed streambed elevations at the bridge. This indicates that the bridge is located above this elevation and would be classified as either tidally influenced or tidally affected. The MLLW elevation is above the surveyed cross section at the bridge sites classified as tidally controlled.

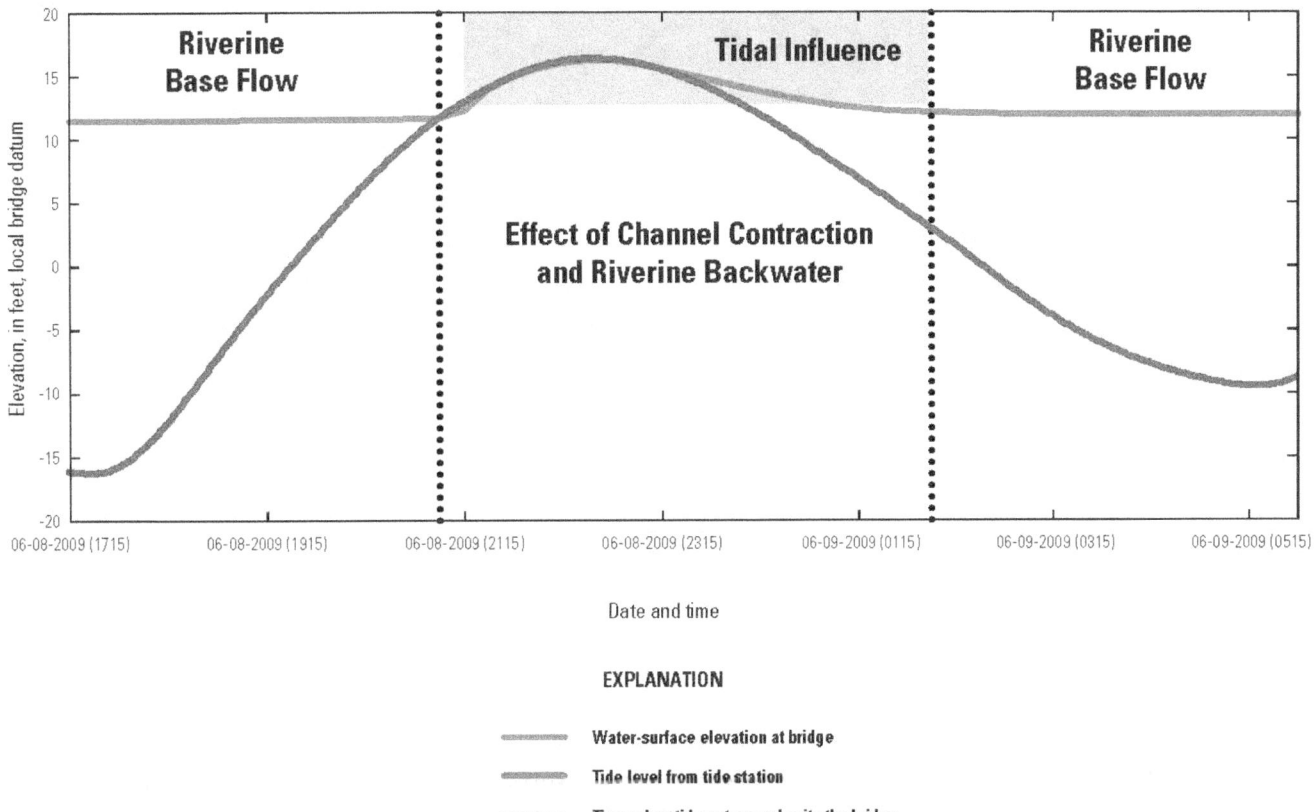

Figure 4. Water-surface elevations from the Sunriset tidal station and at bridge number 634 over the tidally affected Twentymile River, Alaska. Sunrise tidal station elevations have been adjusted to the bridge datum.

Tidal hydrographs are typically symmetrical through a tidal cycle, because the tide rises as fast as it falls. Asymmetrical hydrographs can occur at bridges when tidal flow encounters riverine flow and riverine flow is temporarily affected by backwater caused by the opposing tide. The duration of the outgoing tide is then greater than that of the incoming tide because the outgoing tide includes the release of the backwater-affected riverine component induced by the bridge constriction. The discharge and velocities at the bridge are also greater during the outgoing tide. This situation can lead to scour because the potential for sediment transport through the reach on the outgoing tide is increased due to higher velocities and sustained flow durations that are greater than that of the incoming tide.

The shape of the hydrograph for the tide gage nearest the bridge was compared to the hydrograph measured at the bridge. Tidal analysis from the 33 bridges identified 14 sites where the hydrographs were asymmetric for the outgoing portion of the tide (table 2). The tidally-affected portion of the hydrographs at the other 18 bridges is not always symmetrical,

but asymmetry was not evident in the data collected for this study. An attempt was made to collect data at each bridge during the highest annual tide, but the limited field season and duration of this study precluded data collection during the highest tides at some sites.

The potential for a bridge to act as a constriction during the exchange of tides was also evaluated with aerial imagery. The ratio of the average width of the upstream storage area to the width of the bridge opening was used as a measure of the degree to which a bridge constricts the exchange of tidal and riverine flow (fig. 5). Sites with high ratios of approach channel width to bridge width are susceptible to contraction scour from the increase in flow velocities through the bridge constriction. Bridges in this study were divided into three categories, wherein the width of the approach channel including any overbank area that might be overtopped by tidal or riverine flows is (1) less than or equal to the bridge width, (2) greater than the width of the bridge but less than twice the bridge width, and (3) greater than twice the bridge width (table 2).

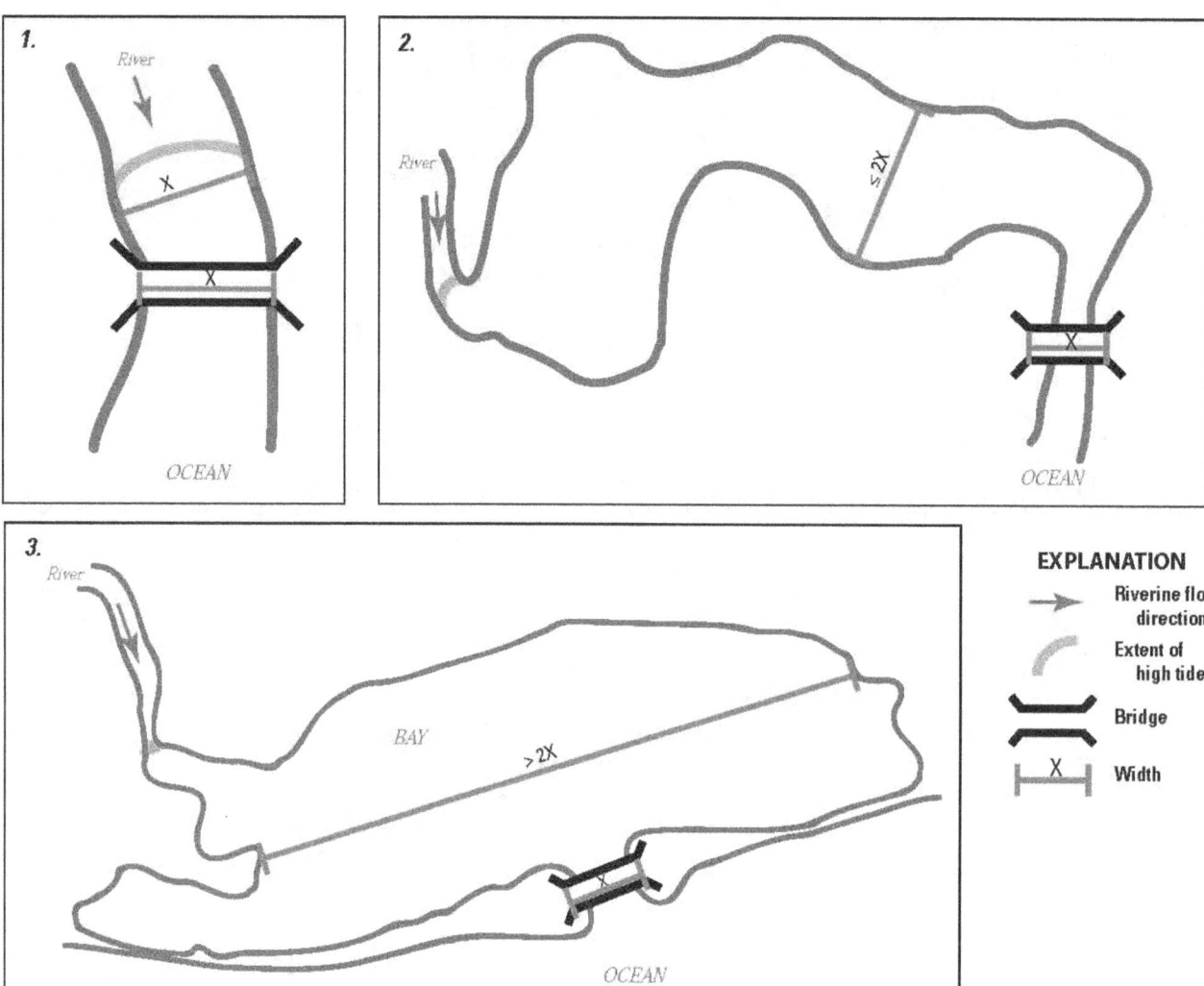

Figure 5. Examples of the three categories used to define the degree to which a bridge can constrict tidal exchanges and backwater riverine flow. Numbers 1–3 indicate the value assigned to the risk posed by upstream storage during a tidal exchange.

Storm Surges

Storm surges occur when coastal waters are forced above the expected high-tide elevation by a combination of low atmospheric pressure and strong winds blowing onshore or along shore where the coast is to the right of the wind direction due to the Coriolis Effect. Areas of Alaska that are particularly susceptible to storm surges are the Bering Sea coast from Bristol Bay to the Bering Strait, Kotzebue Sound, the Chukchi Sea coast, and the Beaufort Sea coast (Wise and others, 1981). Blier and others (1997) analyzed storm-surge induced coastal flooding in Nome, and numerical modeling experiments were conducted for three historical storms. Of the 14 coastal flooding events that occurred from 1900 to 2000 in Nome, all except 2 occurred in the autumn (Blier and others, 1997). Coastal flooding typically ceases from mid-November through late spring because most of the Bering Sea is covered by sea ice. Although many communities are affected by storm surges along the western coast of Alaska, infrastructure is limited and only three bridges in this study are along those shores: ADOT&PF bridge numbers (BN) 1127, 347, and 1385.

Table 2. Hydrograph analysis and ranking of the upstream storage potential at selected bridges over tidal waterways in Alaska.

[N/A, not available, these data were not collected at the bridge]

Bridge No.	Bridge name	Asymmetry in tidal part of the hydrograph	Rank of upstream storage area determined from aerial photographs (1 = low, 3 = high)
214	Swanson River	Yes	2
347	Bonanza Creek	N/A	3
385	Salt Creek	No	3
387	Chilkoot River	Yes	1
399	King Salmon Creek	No	1
400	Leader Creek	No	1
402	Pauls Creek	No	1
429	Blind River	No	2
444	Salmon River	N/A	2
620	Ingram Creek	Yes	1
627	Placer River Overflow	Yes	1
629	Placer River Main Crossing	Yes	1
630	Portage Creek 1	Yes	1
631	Portage Creek 2	Yes	1
634	Twentymile River	Yes	2
636	Peterson Creek	No	1
638	Virgin Creek	Yes	3
639	Glacier Creek	No	1
724	Ketchikan Creek	No	1
989	Sargent Creek	N/A	1
990	Russian River	Yes	1
992	Salonie Creek	N/A	3
1017	Seldovia Slough	Yes	3
1085	Hartney Bay	No	3
1121	Knik River NB	Yes	1
1124	Matanuska River NB	N/A	1
1127	Safety Sound	Yes	3
1149	Kenai River at Kenai Bridge	Yes	3
1197	Lemon Creek NB	No	3
1274	Monashka Creek	No	3
1385	Tununak River	N/A	2
1863	Lemon Creek SB	No	3
2150	Ship Creek	No	2

The U.S. Army Engineer Research and Development Center, Coastal and Hydraulics Laboratory developed frequency-of-occurrence relationships of storm-generated water levels for 17 communities along Kotzebue and Norton Sounds, the Bering Sea, and Bristol Bay (Chapman and others, 2009). Advanced-Circulation model (ADCIRC) simulations were performed for 52 historical storms to develop the return period of storm-surge water levels at each community. The 1-percent AEP water level (AEPWL) prediction for each bridge at risk from storm-surge flooding was based on the AEPWL from the community nearest the bridge (table 3).

Two of the bridges susceptible to storm surges, BN 347 and BN 1127, are east of Nome along the road from Nome to Council (fig. 6). The two bridges cross tidally-controlled inlets to Safety Sound from Norton Sound. Storm surges into Safety Sound are of particular concern because of the large storage area upstream of the bridges and flow restrictions created by the bridge openings. This stretch of coastline is exposed to open waters that extend into the Bering Sea. The two largest storm surges recorded in Nome produced a 10.0 ft rise in sea level on November 10–12, 1974 and 10.5 ft on October 18–20, 2004 (National Climate Data Center, 2011). The 1-percent AEPWL for storm-surge flooding in Nome is 9.7 ft. The October 2004 storm surge damaged the Nome-Council Road a half mile to the east of bridge BN 1127, but no damage was reported at the bridge.

The 1-percent AEPWL for storm-surge flooding at Nome was plotted on cross sections for both bridges (fig. 7) and is only 3.8 ft below the low-beam elevation of the bridge at BN 347. The maximum elevation of woody debris deposited at BN 347 by an undocumented storm that occurred between field visits on September 18, 2008, and September 28, 2010, was surveyed at an elevation only 2.5 ft below the low-beam elevation of the bridge. This debris likely was deposited by wave action rather than a large storm surge because no major storms occurred during that period. Wave action that impinges on a bridge, or wave loading, has the potential to displace a bridge from its foundation. The low-beam elevation of BN 1127 is 8 ft higher than the low-beam elevation of BN 347, so wave loading of the structure is not likely for the 1-percent AEPWL at BN 1127.

Table 3. Bridges at risk from storm surge flooding and the 1-percent annual exceedance probability water level in feet, local datum of the bridges near Nome and Toksook Bay, Alaska.

[The return period of storm-surge water levels at each location was determined from Advanced-Circulation (ADCIRC) model simulations of 52 historical storms (Chapman and others, 2009)]

Bridge No.	Bridge name	Nearest community	1-percent annual exceedance probability surge level (feet, local bridge datum)	Standard deviation	Low elevation of bridge (feet)
1127	Safety Sound	Nome	9.7	1.2	20.5
347	Bonanza Crossing	Nome	9.7	1.2	13.5
1385	Tununak River	Toksook Bay	11.7	1.2	14.5

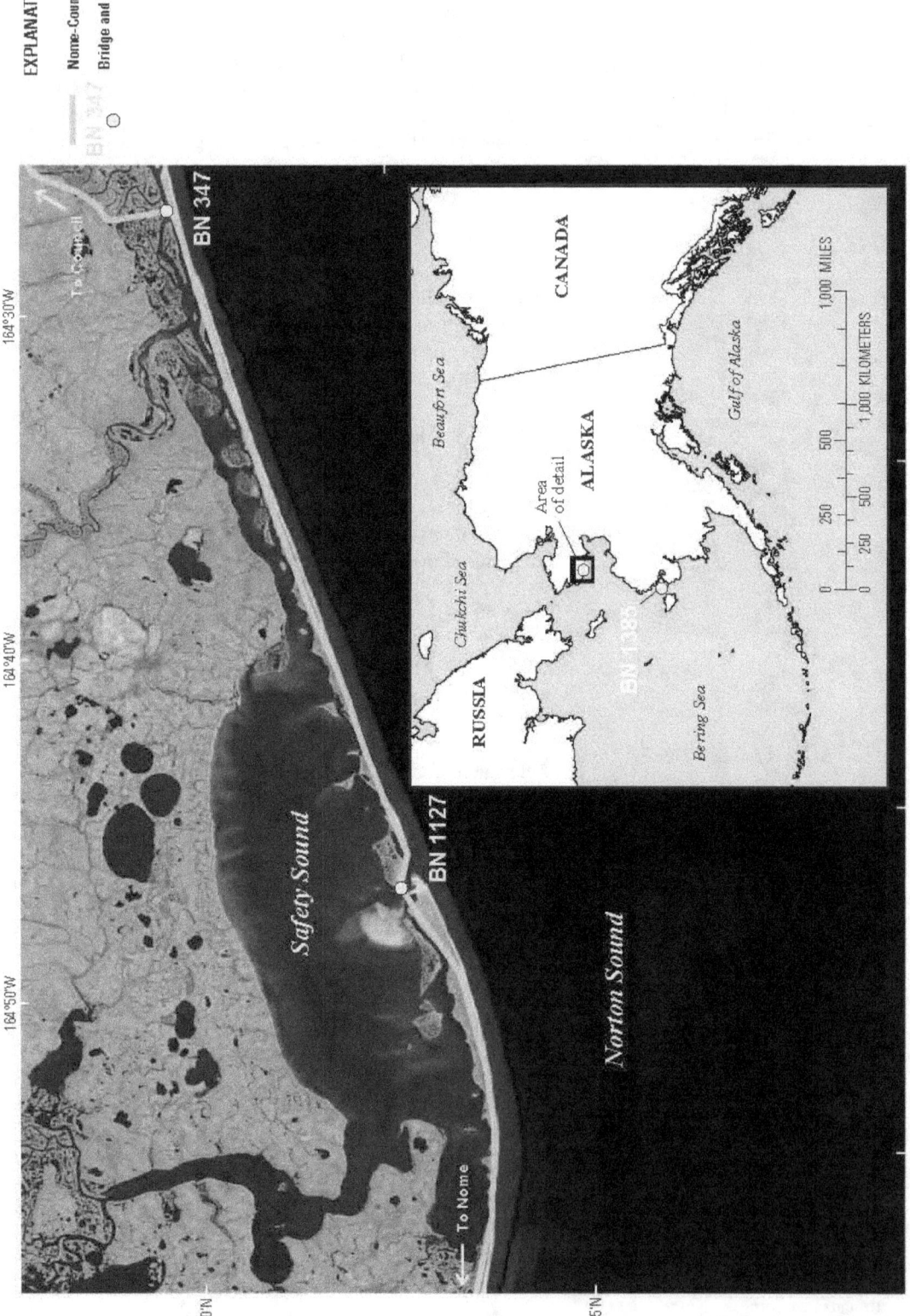

Figure 6.   Location of bridges subject to coastal flooding and detailed location of bridge numbers 347 and 1127 along the Nome–Council Road, Alaska. Background image from NASA's Earth Observatory using data from Moderate Resolution Imaging Spectroradiometer (MODIS).

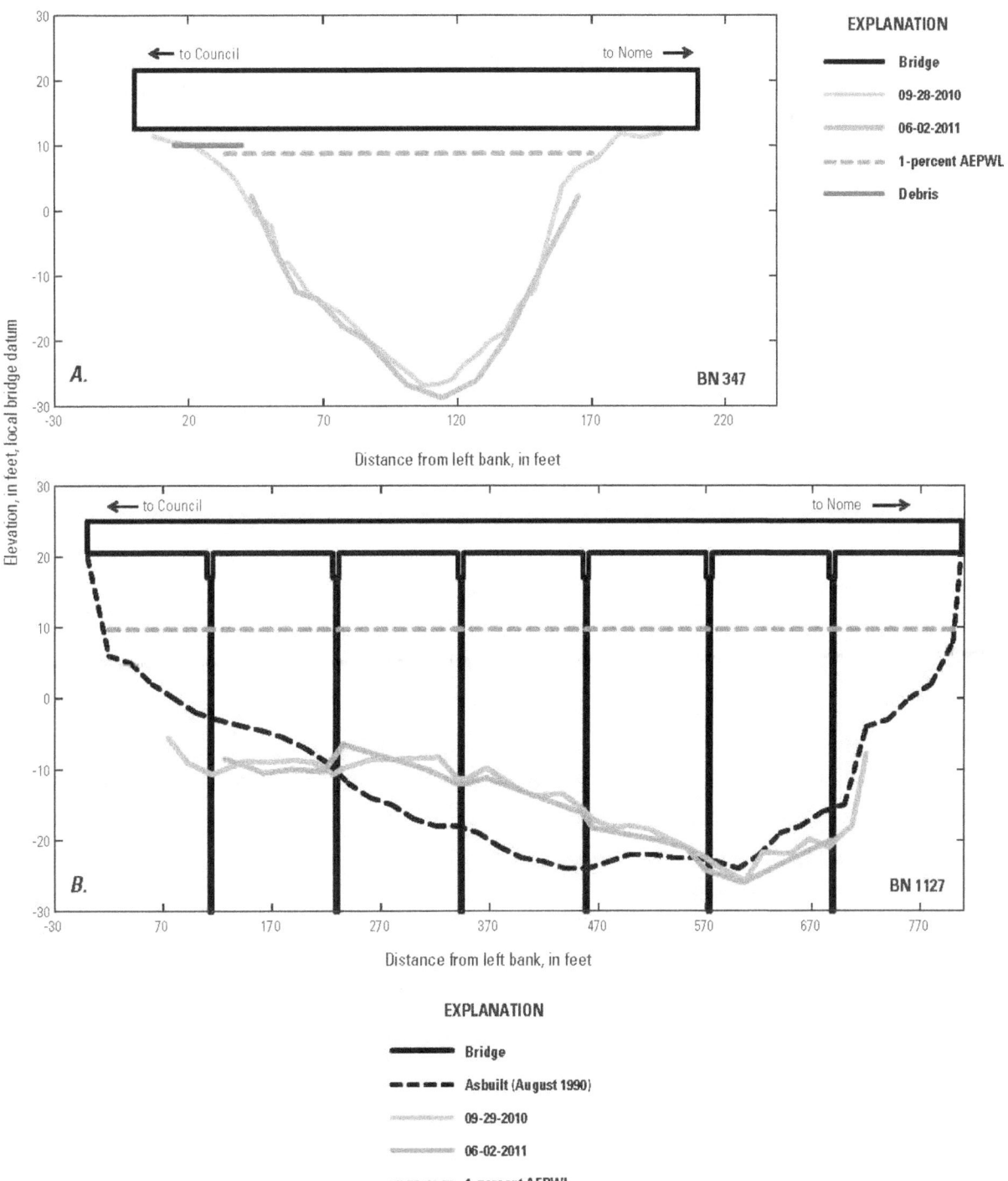

Figure 7. Surveyed channel cross sections at bridges (*A*) BN 347 and (*B*) BN 1127 along the Nome-Council Road showing the 1-percent annual exceedance probability water-surface elevation (AEPWL) for coastal flooding at Nome, Alaska.

The relation between storm-surge elevations at the Nome tide gage and tidal elevations at Safety Sound was studied in autumn 2010. A non-contact stage sensor was installed on BN 1127 to measure water-surface elevations at the bridge at 15-minute intervals from October 2010 until mid-November 2010, when the water began to freeze (fig. 8). The measured water-surface elevations are similar in magnitude and timing to the measured tides at Nome. A storm in late October concurrently increased the water-surface elevations in Nome and at BN 1127 during two tidal cycles, but the water level decreased at a slower rate at the bridge during the second tide cycle. This could result from the temporary storage of river inflows that were affected by backwater caused the high tide, even though rivers at this latitude are usually frozen by late October. Another possibility is that onshore winds backed the tide up at the bridge, but not at the tide gage in Nome.

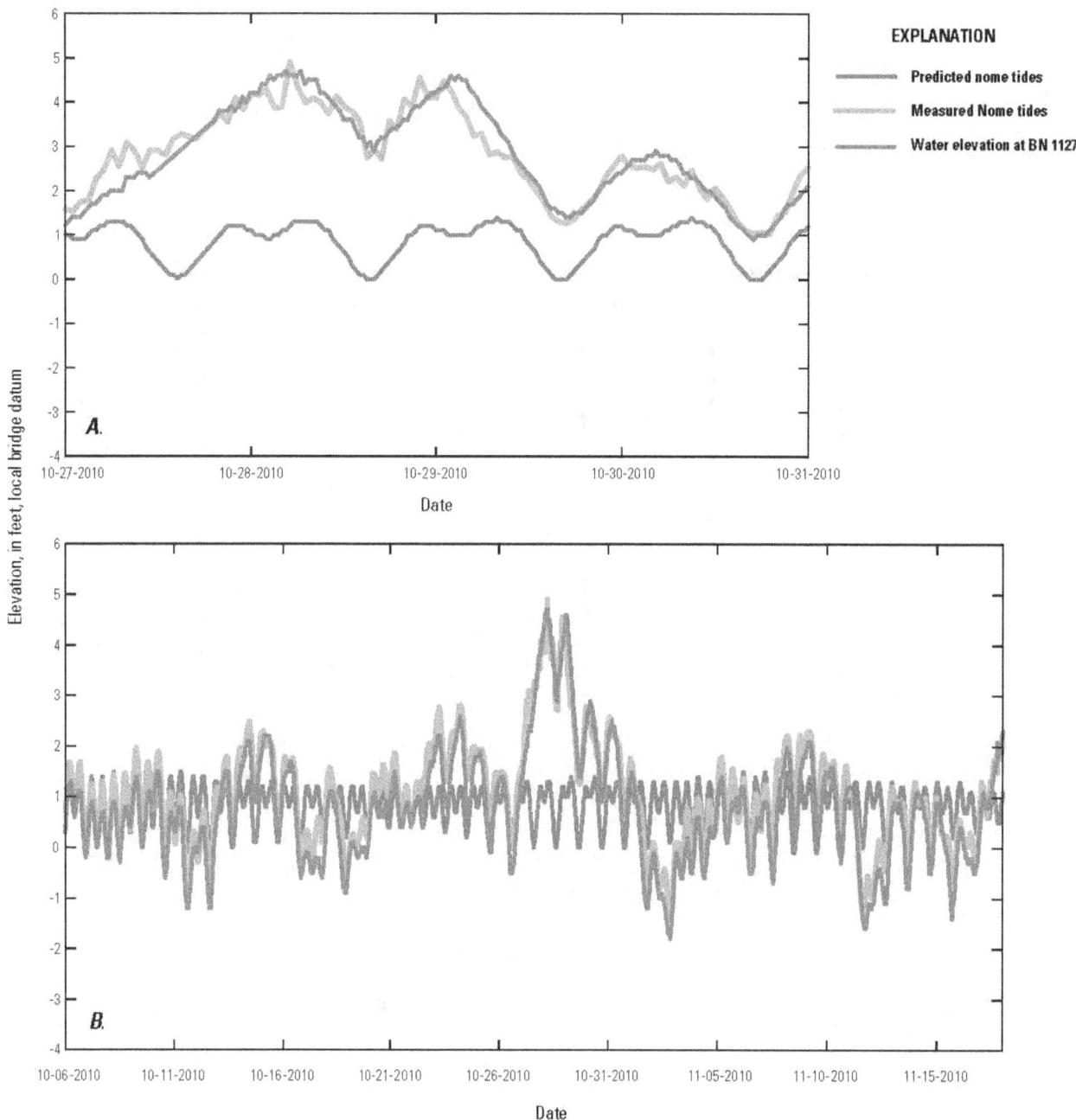

Figure 8. The predicted and measured tides for Nome, Alaska, and the measured water-surface elevation at bridge number (BN) 1127 from (*A*) October 27–31, 2010, and (*B*) October 6–November 17, 2010.

The Tununak River bridge (BN 1385) is approximately 7 mi northwest of the village of Toksook Bay. The bridge is located on a stretch of coastline that faces north into the Bering Sea and is at risk to storm surges. The 1-percent AEPWL for storm-surge flooding in Toksook Bay is 11.7 ft, and this elevation would be 2.8 ft below the lowest elevation of the bridge. The bridge is protected from wave action by a spit, but there is potential for upstream storage during a storm surge. No field measurements of tidal elevations at BN 1385 were made as part of this study. Measurements of water-surface elevations during tidal cycles and storm surges are needed to quantify the impact of storm surges at this bridge.

Velocity Measurements during Tidal Exchanges

Maximum tidal discharge and highest tidal velocities at tidally-controlled and tidally-affected bridges occur during the outgoing tide, which includes any riverine flow that was affected by backwater during the incoming tide. Velocity data were collected during outgoing tides at several tidally-controlled and tidally-affected bridges to (1) determine the scour potential of these flows and (2) compare the magnitudes of tidal and riverine velocities for selected flows at a site. Measurements were made using an acoustic Doppler current profiler (ADCP) tethered either from the downstream side of the bridge or from a manned boat. Tethered boat measurements were stationary and made in the section of the channel where earlier ADCP cross-sectional transects identified the highest velocities. Velocity data were collected from the manned boat along cross sections at the bridge. The highest velocities during a tidal exchange occur at or later than the midpoint between the high and low tides depending on the amount of upstream storage and the degree of contraction caused by the bridge opening (fig. 9). Velocity data were collected during this time frame (the outgoing tide) at 10 bridges, and the peak instantaneous and average velocities were determined (table 4). Most measurements were made after a high tide that was near or exceeded the MHHW. Velocity data also were collected at some of these sites during low tide to compare the magnitude of the riverine velocity to the velocity measured during the tidal exchange. These data measured at low tide were collected only when the flow regime was purely riverine and not affected by the tide.

At sites where both tidal and riverine velocities were measured, the riverine values were always higher, except at BN 1149. The highest velocity during the outgoing tide at BN 1149 was 6.7 ft/s, and the highest riverine velocity measured was 5.1 ft/s. The discharge of the river during the low-tide measurement was 13,100 ft³/s, much lower than the 1-percent AEP discharge of 42,300 ft³/s. Flow velocities at the 1-percent AEP would be higher and are discussed later in the riverine analysis.

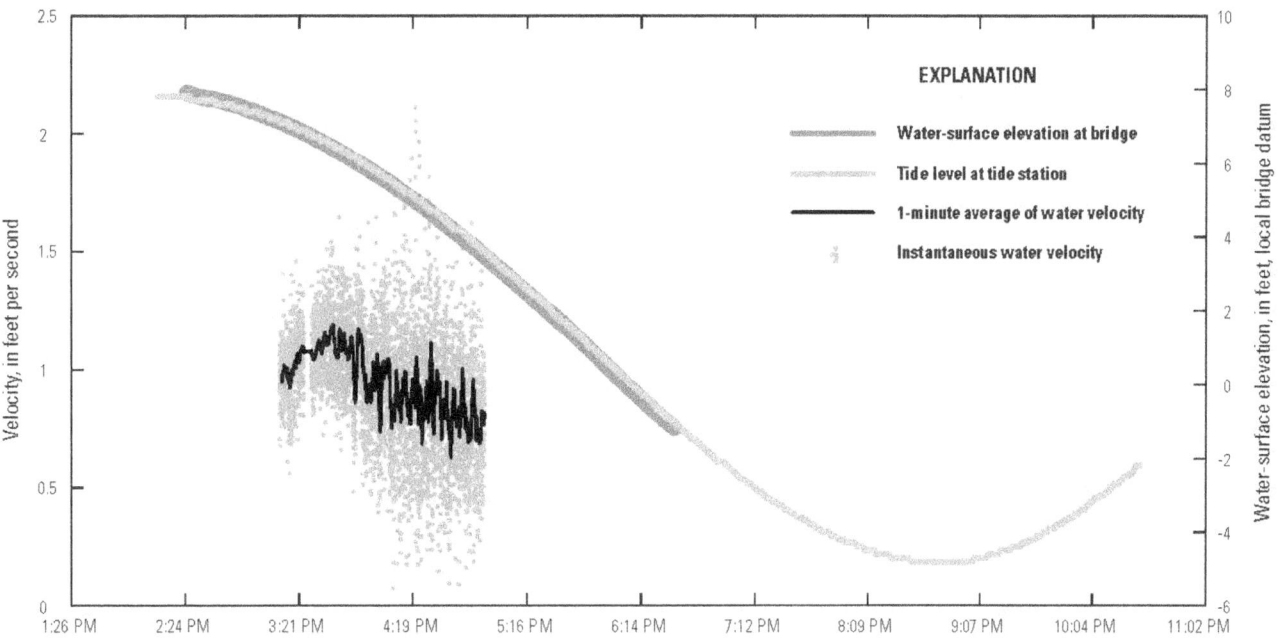

Figure 9. Water-surface elevations and velocity measurements made during a tidal exchange on September 23, 2010, at BN 1085, Hartney Bay, Alaska.

Table 4. Hydraulic measurements made during tidal exchanges at selected bridges over tidal waterways in Alaska.

[Abbreviations: ft, foot; ft/s, foot per second; lb/ft², pound per square foot; ft³/s, cubic foot per second; –, no corresponding riverine data were collected]

Bridge No.	Bridge name	Average depth of flow (ft)	Velocity at tidal exchange (ft/s)		Mannings roughness	Drag coefficient	Estimated boundary shear stress (lb/ft²)	Critical boundary shear stress (lb/ft²)	High-tide elevation at measurement (ft)	Mean high high water elevation (ft)	Measured riverine velocity (ft/s)	Riverine discharge at low tide (ft³/s)
			Maximum	Average								
385	Salt Creek	3.6	2.6	1.1	0.037	0.0125	0.03	0.16	9.6	8.8	–	–
634	Twentymile	4.4	7.0	3.8	0.025	0.0053	0.15	0.16	35.9	33.3	10.1	14,400
638	Virgin Creek	8.5	1.3	0.5	0.032	0.0070	0.00	0.16	35.9	33.3	1.6	17
1017	Seldovia Slough	9.5	5.5	3.8	0.032	0.0068	0.19	0.16	19.7	18.0	5.5	76
1085	Hartney Bay	13.8	2.1	1.1	0.032	0.0060	0.01	0.16	12.0	10.8	2.9	10.9
1127	Safety Sound	19.5	2.8	1.5	0.025	0.0033	0.01	0.16	1.3	2.3	–	–
1149	Kenai at Kenai	19.6	6.7	4.7	0.036	0.0067	0.29	0.16	23.9	20.7	5.1	13,100
1197	Lemon Creek NB	4.3	1.2	0.7	0.030	0.0078	0.01	0.16	19.3	16.3	–	–
1863	Lemon Creek SB	4.3	1.2	0.7	0.030	0.0078	0.01	0.16	19.3	16.3	–	–
2150	Ship Creek	7.0	2.2	0.6	0.032	0.0075	0.01	0.16	28.4	29.2	4.5	67

The ability for tidal or riverine flow to transport sediment is related to the shearing forces of the flow, known as shear stress, that are exerted on the streambed. Sediment transport of noncohesive grains is initiated when the shear stress exceeds the critical shear stress for that grain size. Shear stress cannot be measured directly, but is estimated from measurements of velocity or flow geometry and their relation to bed shear stress (Wilcock, 1996). The bed shear stress was estimated with the quadratic stress law method:

$$\tau_b = \rho\, C_d u^2, \qquad (1)$$

where

τ_b is boundary shear stress, in pounds per square foot $\left(\text{lb/ft}^2\right)$,

ρ is fluid density, in slugs per cubic foot $\left(\text{slugs/ft}^3\right)$,

C_d is the non-dimensional drag coefficient, and

u is vertically averaged flow velocity, in feet per second.

The non-dimensional drag coefficient, C_d, can be related mathematically to the Manning's roughness coefficient with the following equation:

$$C_d = \frac{\left(\dfrac{n_m}{1.515}\right)^2 g}{H^{1/3}}, \qquad (2)$$

where

C_d is the dimensionless drag coefficient,

n_m is the dimensionless Manning's roughness coefficient,

g is the acceleration of gravity, in feet per second, and

H is the mean depth of flow, in feet.

The estimated shear stress from equation 1 was compared to critical shear stress values determined by Julien (1998) for the estimated sediment size at the bridges. Grain-size data are not available for most sites in this study, but Heinrichs and others (2001) determined that 0.033 ft (10 mm) was a reasonable estimate of the median diameter of the bed material for evaluating scour at bridges in Alaska. The critical bed shear stress for a 10 mm particle is 0.16 lb/ft² (Julien, 1998). If the estimated shear stress calculated from the velocity measurements exceeds 0.16 lb/ft², then sediment movement of grain sizes up to 10 mm can be expected. If sediment transport during the outgoing tide is possible, then the tidal exchange is considered a potential scour process at the bridge.

Estimated shear stress values computed from the measured velocities on the outgoing tide exceeded 0.16 lb/ft² at BN 1017 (Seldovia Slough) and BN 1149 (Kenai River at Kenai) (table 4). The sensitivity of this computation to grain size was evaluated by reducing the estimated grain size diameter to 0.016 ft (5 mm). This reduced the critical boundary shear stress to 0.07 lb/ft² and then BN 634 would be included in the sites where the estimated shear-stress value exceeded the critical shear-stress value. The risk of scour at BN 1017 is minimal because both bridge piers are founded on bedrock. Scour during tidal exchanges is primarily a concern at sites where sediment transport occurs only during the outgoing tide and there is no sediment replenishment from riverine inflow. Bedload sediment transport was noted in the ADCP data collected during riverine discharges at BN 634 and BN 1149, indicating that any scour during an outgoing tide would be limited by infilling during subsequent riverine input.

Riverine Analysis

The majority of the bridges studied were characterized as either tidally affected or tidally influenced (table 1), and the primary risk for streambed scour at these sites is from riverine flows rather than tidal fluctuations. Streambed scour was evaluated for these bridges using guidelines from HEC-18 (Richardson and Davis, 2001) and previous scour studies in Alaska by Heinrichs and others (2001) and Conaway (2004). Hydraulic variables were computed for the 1-percent AEP riverine discharge with the Hydrologic Engineering Center River Analysis System (HEC-RAS; Brunner, 2010).

One-dimensional steady-state models of the 1-percent AEP discharge were constructed using a combination of existing data and data gathered as part of this study. Existing data consisted primarily of the bridge as-built plans provided by ADOT&PF. The as-built plans generally included a detailed topographic map of the channel near the bridge at the time of construction and limited hydrologic information. Other sources of existing data include discharge measurements, streamflow-gaging station records, or ADOT&PF bridge inspection reports. The 1-percent AEP discharge was either obtained from the as-built plans or it was estimated using regional regression equations as outlined by Curran and others (2003).

Estimates of scour were computed using the equations and methodology outlined in HEC-18 (Richardson and Davis, 2001). Contraction scour was computed at all sites and pier scour was computed at crossings supported by piers. Abutment scour was not evaluated because of the large computational uncertainties and because most abutments on bridges in Alaska are armored with riprap to inhibit scour (Heinrichs and others, 2001). Flow widths, depths, and velocities were calculated with HEC-RAS and were used to compute the scour estimates for the 1-percent AEP discharge.

Data Collection and Model Parameters

Hydraulic models were constructed using a combination of field data and the bridge as-built plans. Bridge soundings or discharge-measurement data, in combination with geometry from the bridge as-built plans, were used to construct cross sections at the upstream and downstream face of the bridge. Channel-geometry data from the cross section at the bridge were used with overbank data from bridge as-built plans or a field survey to construct full-valley cross sections of the approach and exit sections associated with the bridge. Field data collected at a bridge for the riverine analysis typically included a discharge measurement, channel soundings from the bridge deck, water-surface slope through the reach, flow angle of attack on the piers, geomorphic observations, and photographs. Full valley cross sections were surveyed at sites at which topographic data included in the as-built plans were limited. Stream discharge was not measured at some sites because the discharge at the time of the site visit was insufficient to calibrate a hydraulic model of the 1-percent AEP discharge.

Channel roughness coefficients for each site were computed by the Manning equation using slope and discharge-measurement data obtained for this study (Hicks and Mason, 1991). If these data were not available, roughness values of 0.035 for the channel and 0.045 for the overbanks were assumed for the model. These default values were considered to be a good approximation for streams in Alaska (Heinrichs and others, 2001). When photographs were available, the overbank roughness coefficients were estimated from inspection of the photographs in combination with interpretive techniques outlined by Hicks and Mason (1991).

Water-surface profiles were computed with the HEC-RAS model for discharges that were measured and for the computed 1-percent AEP discharges. Models were calibrated for sites where discharge and corresponding water-surface elevation data were collected. In those instances, channel-roughness values were adjusted until agreement was reached between measured and modeled water-surface elevation and velocity. Data used in the determination of the downstream boundary condition for modeling hydraulic conditions varied and either was a known water-surface elevation, a surveyed water-surface slope, or a slope obtained from a topographic map. The downstream boundary condition for modeling hydraulic conditions associated with the 1-percent AEP discharge was either the energy-gradient slope computed from the calibration discharge or the water-surface elevation for a low tide. The low-tide elevation was used at sites where the downstream-most cross section was located in the intertidal zone. A high-flow low-tide scenario represents the hydraulic conditions with the steepest water-surface slope and highest velocities and is therefore the most conservative estimate of scour at a tidally-affected and tidally-influenced river crossing.

If Froude numbers exceeded 1.0 for any cross section, an upstream boundary condition (the slope of the energy gradient, determined from the calibration discharge) was included and the model was rerun as a mixed-flow regime. Interpolated cross sections were inserted between surveyed cross sections if the water-surface profile varied greatly over a short reach, the channel expanded or contracted rapidly, or the channel gradient changed abruptly. Hydraulic variables used to construct the models are summarized in table 5.

The use of hydraulic models poses several limitations and introduces several sources of inaccuracies. Hydraulic models in this study were constructed and calibrated on the basis of existing information and data collected at the time of the field visits. The calibrated models then were extrapolated to accommodate the 1-percent AEP flood flow. Simulations of flood flows with models calibrated to smaller discharges can result in hydraulic inaccuracies. Channel roughness values can vary with stage as the influence of channel features changes with depth. Calibrating channel roughness to low discharges and then extrapolating to higher discharges can be an overestimation because channel roughness normally decreases with increasing depth. In many cases, flow in the overbank areas, where the roughness is generally higher, compensates for the decreased roughness in the channel that results from the greater flow depths (Conaway, 2004). One-dimensional models capture a small part of the active processes in the channel, but are efficient at making predictions over long length and time scales (Nelson and others, 2003). For many of the sites in this study, a multi-dimensional flow model would better capture the complex hydraulics associated with flood flows at bridges. Data requirements for these types of models, however, were prohibitive for this level of study. The hydraulic variables generated by HEC-RAS for complex flow conditions are considered the best available for this level of study, and multi-dimensional models are recommended for sites with complex flow regimes. Several models were calibrated with hydraulic variables collected at discharges that were much lower than the expected magnitude for the 1 percent AEP flow. This can result in a poor calibration and affect the computed scour values for the 1-percent AEP discharge. Additional hydraulic data collected at higher discharges is recommended for the locations with high scour estimates and calibration data collected at low flows.

Estimation of Contraction and Pier Scour

Estimates of contraction and pier scour for the 1-percent AEP discharge were computed from the HEC-RAS modeling results using the methods and equations outlined in HEC-18 (Richardson and Davis, 2001). The HEC-18 equations were chosen so that the methodology would be consistent with previous evaluations of streambed scour in Alaska. Contraction scour was computed for all sites and pier scour was computed for bridges supported by piers.

Table 5.   Selected hydraulic variables used to construct hydraulic models for the analyses of streambed scour at selected bridges over tidal waterways in Alaska.

[**Abbreviations:** AEP, annual exceedance probability; ft^3/s, cubic foot per second]

Bridge No.	Bridge name	Manning's roughness coefficient		Water-surface slope	Source of water-surface slope	Discharge for calibration (ft^3/s)	1-percent AEP discharge (ft^3/s)
		Channel	Overbank				
214	Swanson River	0.035	0.045	0.0022	Survey	45.2	3,040
385	Salt Creek	0.037	0.045	0.0025	Survey	392	578
387	Chilkoot River	0.032	0.050	0.0010	Survey	185	12,200
399	King Salmon Creek	0.032	0.045	0.0016	Survey	590	3,120
400	Leader Creek	0.032	0.045	0.0139	Survey	5.99	100
402	Pauls Creek	0.032	0.045	0.0018	Survey	350	1,700
418	Sheep Creek	0.040	0.045	0.0050	Survey	62	1,140
429	Blind River	0.030	0.045	0.0017	Survey	67	2,710
444	Salmon River	0.032	0.045	0.0003	As-built	119	10,100
620	Ingram Creek	0.028	0.045	0.0020	Survey	373	5,300
627	Placer River Overflow	0.018	0.045	0.0008	Survey	4,710	10,200
629	Placer River Main	0.026	0.045	0.0004	Survey	4,480	10,200
630	Portage #1	0.020	0.045	0.0004	Survey	1,000	7,160
631	Portage #2	0.025	0.045	0.0004	Survey	5,990	7,160
634	Twentymile	0.025	0.045	0.0004	Survey	8,677	18,420
636	Peterson Creek	0.040	0.040	0.0058	Survey	20.3	547
638	Virgin Creek	0.032	0.045	0.0090	Survey	17	563
639	Glacier Creek	0.032	0.045	0.0026	Survey	309	4,920
724	Ketchikan Creek	0.040	0.045	0.0046	Survey	821	3,180
732	Gold Creek	0.055	0.060	0.0300	As-built	23	3,250
787	Salmon Creek at Twin Lakes	0.041	0.045	0.0084	Survey	38	3,290
989	Sargent Creek	0.022	0.045	0.0022	Survey	70	4,230
990	Russian River	0.034	0.045	0.0003	Survey	706	4,730
992	Salonie Creek	0.040	0.044	0.0005	Survey	151	5,090
1017	Seldovia Slough	0.032	0.045	0.0360	Survey	76	200
1085	Hartney Bay	0.032	0.045	0.0035	Survey	10	2,810
1121	Knik River NB	0.023	0.045	0.0009	Survey	73,600	120,000
1124	Matanuska NB	0.028	0.045	0.0001	Survey	37.1	47,000
1149	Kenai at Kenai	0.036	0.045	0.0005	Survey	13,100	42,300
1188	Salmon Creek at Egan Drive	0.041	0.045	0.0084	Survey	38	3,290
1197	Lemon Creek NB	0.030	0.045	0.0020	Survey	128	6,940
1274	Monashka Creek	0.035	0.045	0.0030	Survey	59	2,210
1783	Spruce Creek	0.035	0.045	0.0090	Survey	323	3,930
1863	Lemon Creek SB	0.030	0.045	0.0020	Survey	128	6,940
2150	Ship Creek	0.032	0.045	0.0100	Survey	170	1,900

Contraction Scour

Contraction scour was estimated using Laursen's (1963) live-bed contraction-scour equation (eq. 3). The live-bed equation was selected because active bedload transport is occurring at all of the sites being considered in this study. The equation is based on the understanding that scour reaches a maximum when sediment transport into the contracted section (bridge cross section) equals sediment transport out of the contracted section or when the mean flow velocity equals the critical velocity of the mean-diameter bed material. For all sites studied, the bed material was assumed to be mobile because bed-material size was generally classified as fine gravel or smaller and, therefore, would be expected to be mobile during periods of high flow. The equation recommended in HEC-18 (Richardson and Davis, 2001) for estimating live-bed contraction scour is

$$y_s = y_1 \left[\left(\frac{Q_2}{Q_1} \right)^{\frac{6}{7}} \left(\frac{W_1}{W_2} \right)^{k_1} \right] - y_0, \qquad (3)$$

where

y_s is the contraction scour depth, in feet;

y_1 is the average depth in the upstream main channel, in feet;

y_0 is the average depth in the contracted section before scour, in feet;

Q_1 is the discharge in the main channel of the approach section that is transporting sediment, in cubic feet per second;

Q_2 is the discharge in the contracted section, in cubic feet per second;

W_1 is the width of the main channel of the approach section that is transporting sediment, in feet;

W_2 is the width of the main channel in the contracted section that is transporting sediment, in feet; and

k_1 is a coefficient that accounts for the predominant sediment transport condition. For this study, transport at all sites was assumed to be mostly through bed-material discharge; the coefficient for this condition is 0.59.

Determining the depth in the contracted section before scour (y_0) is difficult because the current channel geometry in the bridge section has likely been modified by contraction scour. For this study and the previous USGS studies on streambed scour in Alaska (Heinrichs and others, 2001, Conaway, 2004), the depth in the approach channel (y_1) was substituted for y_0 in equation 2. If y_1 is measured far enough away from the bridge so that the channel geometry is not affected by the downstream contraction, it is a good approximation of the average depth in the contracted section before scour (y_0). Estimated contraction-scour depths and hydraulic variables used in the computation for the 1-percent AEP discharge are presented in table 6.

Pier Scour

Local scour at bridge piers was evaluated with the techniques and equation outlined in HEC-18. This empirically derived equation accounts for the shape and dimensions of the pier, flow depth and velocity, the flow angle of attack on the pier, and the size of the bed material. The HEC-18 equation for estimating pier scour is

$$y_s = 2.0 \, y_0 K_1 K_2 K_3 K_4 \left(\frac{a}{y_0} \right)^{0.65} Fr_1^{0.43}, \qquad (4)$$

where

y_s is pier-scour depth, in feet;

K_1 is the correction factor for pier-nose shape, dimensionless;

K_2 is the correction factor for flow angle of attack on the pier, dimensionless;

K_3 is the correction factor for channel bedform, dimensionless;

K_4 is the correction factor for armoring of the streambed, dimensionless;

a is the pier width, in feet;

y_0 is the flow depth directly upstream of the pier in feet; and

Fr_1 is the Froude number just upstream from the pier, dimensionless.

The correction factor for channel bedform, K_3, is 1.1 for bedforms with a magnitude less than 10 ft, which can be assumed for all sites in this study. The correction factor for armoring of the bed material, K_4, was not used in this analysis or in previous USGS studies of streambed scour at bridges in Alaska because of paucity of data on bed-material size and gradation. Estimated pier-scour depths and hydraulic variables used in the computation for the 1-percent AEP discharge are presented in table 7.

Table 6.   Computed contraction-scour depths, and hydraulic variables used in computation for the 1-percent annual exceedance probability (AEP) discharge at selected bridges over tidal waterways in Alaska.

[**Abbreviations:** ft^3/s, cubic foot per second; ft, foot]

Bridge No.	Bridge name	Discharge at bridge (ft^3/s)	Width of approach channel (ft)	Discharge at approach (ft^3/s)	Flow depth in approach (ft)	Width of channel at bridge (ft)	Depth of flow at bridge (ft)	Depth of contraction scour (ft)
214	Swanson River	3,040	120	3,040	6.2	106.8	5.7	0.4
385	Salt Creek	578	138	578	2.0	95.7	2.8	0.5
387	Chilkoot River	12,200	171	12,200	7.9	163.3	9.9	0.2
399	King Salmon Creek	3,120	116	3,120	5.7	93.0	6.9	1.2
400	Leader Creek	100	11	100	1.6	10.9	0.8	0.0
402	Pauls Creek	1,700	117	1,700	4.7	73.5	5.0	1.2
418	Sheep Creek	1,140	96	1,140	1.2	43.3	2.8	0.7
429	Blind River	2,710	204	2,710	7.4	117.6	7.2	2.9
444	Salmon River	10,100	118	8,963	12.7	149.0	10.9	1.4
620	Ingram Creek	5,300	73	4,930	7.5	141.0	5.2	0.5
627	Placer River Overflow	10,200	290	10,159	5.1	234.6	6.1	0.7
629	Placer River Main	10,200	452	10,200	5.2	413.5	5.1	0.3
630	Portage number 1	7,160	132	7,072	7.0	155.4	5.8	0.1
631	Portage number 2	7,160	151	7,105	6.6	212.6	9.0	0.0
634	Twentymile	18,420	481	18,420	11.3	483.4	7.6	0.0
636	Peterson Creek	547	31	547	1.8	66.6	1.3	0.0
638	Virgin Creek	563	71	563	3.1	28.3	2.9	2.2
639	Glacier Creek	4,920	115	4,920	4.6	120.9	5.0	0.0
724	Ketchikan Creek	3,180	42	3,180	6.3	63.0	6.8	0.0
732	Gold Creek	3,250	70	3,250	5.0	62.2	5.3	0.4
787	Salmon Creek at Twin Lakes	3,290	95	3,172	4.0	78.4	4.6	0.6
989	Sargent Creek	4,230	72	4,177	6.3	64.0	5.1	0.5
990	Russian River	4,730	106	4,618	6.8	99.4	5.6	0.4
992	Salonie Creek	5,090	83	3,710	9.2	90.9	8.5	2.7
1017	Seldovia Slough	200	101	200	0.7	53.6	0.8	0.3
1085	Hartney Bay	2,810	213	2,809	3.9	62.5	5.1	4.1
1121	Knik River northbound	120,000	1,408	120,000	11.9	1,411.9	13.6	-0.0
1124	Matanuska northbound	47,000	1,037	47,000	6.7	1,053.6	8.8	0.0
1149	Kenai at Kenai	42,300	752	42,300	10.0	715.7	9.9	0.3
1188	Salmon Creek at Egan	3,290	51	2,865	7.2	76.2	5.9	0.9
1197	Lemon Creek northbound	6,940	155	6,265	9.4	134.6	8.4	1.7
1274	Monashka Creek	2,210	73	2,038	5.1	92.0	4.5	0.4
1783	Spruce Creek	3,930	154	3,930	4.8	66.0	6.6	3.1
1863	Lemon Creek SB	6,940	155	6,265	9.4	133.8	8.4	1.8
2150	Ship Creek	1,900	54	1,900	3.6	47.5	3.68	0.3

Evaluation of Streambed Scour Computations

Contraction and pier-scour depths were computed for the 1-percent AEP discharge at the 35 bridges at risk of scour from riverine flows. Contraction scour exceeded 2 ft at 5 bridges, with a maximum of 4.1 ft at BN 1085 (table 6). Pier scour was at least 2 ft at 29 bridges and at least 4 ft at 15 bridges (table 7). The validity of the scour computations depends on (1) the quality and quantity of data used to develop the hydraulic models, (2) the ability of the model to accurately extrapolate hydraulic parameters from the calibrated condition to the 1-percent AEP discharge, and (3) the accuracy of the predictive equations. The amount of data available and collected for this study differed from site to site, but was sufficient to develop hydraulic models and determine the susceptibility of each bridge to contraction and pier scour.

Table 7. Computed pier-scour depths and hydraulic variables used in computation for the 1-percent annual exceedance probability (AEP) discharge at selected bridges over tidal waterways in Alaska.

Bridge No.	Bridge name	Approach flow depth at bridge (feet)	Froude No. at bridge	Pier nose shape	Flow angle of attack (degrees)	Correction factor for flow angle of attack (K_2 coefficient)	Pier width (feet)	Pier length (feet)	Approach flow depth at pier (feet)	Computed depth of pier scour (feet)	Minimum bed elevation at bridge (feet)	1 percent AEP water-surface elevation (feet)
214	Swanson River	5.7	0.35	round	0	1.0	2.0	34	5.7	4.0	2.8	13.8
385	Salt Creek	2.8	0.21	group of cylinders	0	1.0	1.0	27	2.9	1.6	83.8	87.9
387	Chilkoot River	9.9	0.40	square	0	1.0	1.0	22	10.0	3.7	5.0	17.7
399	King Salmon Creek	6.9	0.31	square	0	1.0	1.0	24	7.0	2.9	-3.0	6.4
400	Leader Creek	0.8	0.90	group of cylinders	0	1.0	1.0	25	0.9	2.0	5.0	6.7
402	Pauls Creek	5.0	0.34	square	0	1.0	1.0	25	4.9	2.7	-1.0	6.1
418	Sheep Creek	2.8	1.00	group of cylinders	0	1.0	1.0	31	2.8	3.2	11.6	15.2
429	Blind River	7.2	0.21	group of cylinders	0	1.0	0.8	16	7.2	1.9	2.4	10.5
444	Salmon River	10.9	0.33	group of cylinders	10	2.1	1.0	25	10.9	6.5	-5.5	10.9
620	Ingram Creek	5.2	0.52	sharp	0	1.0	1.5	40	7.7	4.0	12.7	22.1
627	Placer River Overflow	6.1	0.46	sharp	20	2.9	2.7	35	6.6	16.5	4.4	17.2
629	Placer River Main	5.1	0.38	sharp	0	1.0	2.7	35	5.3	4.4	8.4	17.2
630	Portage #1	5.8	0.57	sharp	0	1.0	1.5	32	6.2	3.8	12.0	20.6
631	Portage #2	9.0	0.22	sharp	0	1.0	1.5	35	9.2	2.9	2.8	17.3
634	Twentymile	7.6	0.31	round	0	1.0	1.5	35	7.6	3.5	-1.0	15.5
636	Peterson Creek	1.3	1.00	sharp	0	1.0	1.5	30	1.5	3.1	16.0	18.8
638	Virgin Creek	2.9	0.73	sharp	15	2.5	2.7	35	3.0	4.8	7.6	11.1
639	Glacier Creek	5.0	0.68	square	0	1.0	2.5	33	5.0	6.5	11.5	17.8
724	Ketchikan Creek	6.8	0.50	none	-	-	-	-	-	-	-10.2	1.5
732	Gold Creek	5.3	0.68	round	0	1.0	1.5	88	5.4	4.3	3.1	11.0
787	Salmon Creek at Twin Lakes	4.6	0.75	none	-	-	-	-	-	-	15.9	21.7
989	Sargent Creek	5.1	0.80	group of cylinders	0	1.0	1.2	22	5.8	4.1	2.0	11.7
990	Russian River	5.6	0.58	group of cylinders	0	1.0	1.2	29	5.8	3.6	3.8	11.7
992	Salonie Creek	8.5	0.36	group of cylinders	0	1.0	1.3	29	9.8	3.6	-0.9	13.5
1017	Seldovia Slough	0.8	0.89	group of cylinders	0	1.0	2.0	34	0.8	3.0	3.9	5.5
1085	Hartney Bay	5.1	0.70	group of cylinders	0	1.0	1.0	25	5.2	3.4	-7.8	1.9
1121	Knik River NB	13.6	0.30	round	0	1.0	4.8	40	13.9	9.1	3.7	22.6
1124	Matanuska NB	8.8	0.30	round	0	1.0	4.3	40	9.0	7.2	12.8	27.5
1149	Kenai at Kenai	9.9	0.34	sharp	0	1.0	4.3	44	9.9	8.0	-16.3	-0.6
1188	Salmon Creek at Egan	5.9	0.51	group of cylinders	0	1.0	1.5	109	7.3	4.3	10.0	17.8
1197	Lemon Creek NB	8.4	0.37	group of cylinders	0	1.0	2.0	40	9.5	2.0	8.8	18.8
1274	Monashka Creek	4.5	0.42	group of cylinders	0	1.0	1.0	35	5.3	2.7	0.3	7.3
1783	Spruce Creek	6.6	0.62	None	-	-	-	-	-	-	19.7	27.0
1863	Lemon Creek SB	8.4	0.37	group of cylinders	0	1.0	2.0	40	9.4	2.0	8.2	18.6
2150	Ship Creek	3.68	0.66	group of cylinders	0	1	3	43	4.42	6.3	-5.1	1.9

Hydraulic models are sensitive to channel roughness coefficients used in the water-surface profile computations. If the channel roughness is overestimated (too high), model simulations will indicate higher water-surface elevations, lower water velocities, and greater flow areas than would actually occur. Ideally, the channel roughness coefficient is computed using the Manning equation and data from a discharge measurement that approximates the flow magnitude of interest, and is verified with surveyed water-surface elevations. If a discharge measurement of sufficient magnitude and surveyed water-surface elevations were not available, then a default value of 0.035 was used for the channel roughness coefficient. The default value also was used if the measured discharge was low enough so that the flow conditions would not be representative of the flood conditions. Varying the channel roughness coefficient has an equal effect on all the variables used in the computation of contraction scour, and precision is only critical when the width of the approach channel varies with slight changes in water-surface elevation (Conaway, 2004).

Computations of pier scour are more sensitive to channel roughness than contraction scour because changes in channel roughness affect the flow depth upstream of the pier, the approach velocity, and the associated Froude number. The other variables in the pier-scour equation are unaffected. Sensitivity analysis of the pier-scour equation indicated that the approach velocity's influence on pier scour is second only to pier width, which is a constant (Glenn, 1994). Conaway (2004) determined that a decrease in channel roughness from 0.080 to 0.035 resulted in a 23 percent increase in computed pier scour at one location. Further analysis is recommended for sites that were identified as having unstable channels (table 8), have piers with a shallow foundation, and had estimated rather than computed channel roughness values.

Repeated Cross-Section Surveys

Repeated cross-section surveys can provide information on long-term aggradation and degradation or stability of a stream channel. Biennial bridge inspections made by ADOT&PF usually include a survey of a cross section at the upstream face of the bridge. Changes in bed elevation between surveys (table 5) indicate the relative mobility of the streambed and whether scour, fill, or both, are taking place. Streambed elevation changes also can indicate changes in sediment supply, channel migration, and littoral drift.

Continued decreases in streambed elevations indicate long-term degradation, which is of particular concern at bridges over tidal waterways because an increase in channel area results in a higher volume of water being exchanged through the bridge during tidal cycles and thus higher water velocities.

Repeated cross-sectional surveys made by ADOT&PF or USGS were obtained for 41 bridges and were analyzed for channel instability (appendix B). Channel soundings also were collected by the USGS for most of the sites in this study from the upstream and/or downstream side of the bridge deck with sounding weights. Data collected by ADOT&PF were synthesized with data collected for this study, and cross sections were analyzed visually for any change in shape or elevation. Vertical differences less than 2 ft over multiple surveys were considered to be an indication of negligible channel instability because measurement techniques differed between ADOT&PF and USGS. None of the repeated surveys indicated long-term channel degradation, but 14 sites showed signs of unstable channels (table 8).

Table 8. Sites where channel instability was identified from repeated cross-sectional surveys at the bridge.

Bridge No.	Bridge name	Relative bed elevation change between surveys (feet)
387	Chilkoot River	2.1
444	Salmon River	3.7
627	Placer River Overflow	4.9
629	Placer River Main	4.2
630	Portage Creek 1	3.9
631	Portage Creek 2	6.1
634	Twentymile River	4.3
638	Virgin Creek	4.9
639	Glacier Creek	2.6
990	Russian River	4.8
1121	Knik River	6.1
1124	Matanuska River NB	7.7
1149	Kenai River at Kenai	6.6
2150	Ship Creek	3.0

Overall Risk to Bridges over Tidal Waterways

Bridges over tidal waterways are subject to a number of processes that can threaten the stability of the structure. In some instances, an individual process dominates. For example, bridges over steep mountain streams that were classified (in this study) as tidally influenced are likely at risk of scour only from riverine flow. An analysis of risk is more complex for sites that are subject to several processes that individually do not pose a significant threat to a bridge, but their combined and additive effects do. A bridge that was classified as tidally affected with upstream storage area and at risk of riverine flooding could potentially experience a flood in conjunction with a high tide. The combination of these risks could then produce potentially hazardous hydraulic conditions at the bridge. Each individual risk that was evaluated was assigned a value from 0 to 3 to rank the potential risk at each location. A risk value of 0 would indicate no risk while a risk value of 3 would indicate high risk.

Only three sites considered in this study were identified as being subject to storm surges, and all three were assigned a risk value of 3. The ranking of the sites where tidal velocities were measured was based upon the magnitude of the tidal velocity compared to the computed critical velocity for the estimated bed material size. A risk value of 3 was assigned to the sites where the maximum measured velocity exceeded the critical velocity, a value of 2 was assigned if the measured velocity was greater than half the critical velocity, and a value of 1 was assigned if the measured velocity was less than half the critical velocity. The range of values and the assigned risk value are summarized in table 9.

Risk values assigned to the upstream storage potential are based on visual interpretation of aerial photos and are therefore more subjective than the risk values assigned to computed scour depths. Bridges that are tidally influenced were all assigned a risk value of 0. Tidally controlled and affected bridges were assigned a risk value based upon the geometry of the channel upstream of the bridge. A risk value of 1 was assigned to sites where the tidal exchange was confined to the width of the bridge opening. A risk value of 2 was assigned to sites where the lateral extent of the tidal flood upstream of the bridge was greater than the bridge opening, but less than twice the width of the bridge opening. Sites where the lateral extent of the tidal flood upstream of the bridge was greater than twice the width of the bridge opening were assigned a value of 3.

Scour risk associated with channel instability was ranked on the basis of the measured range of bed elevations determined from repeated channel surveys. Values assigned to assess the individual risks from contraction and pier scour were based upon the range of computed values from the riverine analyses and are summarized in table 9. A risk value of 0 was assigned if no riverine analysis was made because the site was tidally controlled, the computed scour value was actually 0, or if no pier-scour computations were made because the bridge did not have piers. The computed scour depths were used to determine the risk values before considering any mitigating factors such as shallow scour-hole depths relative to deep pile-tip elevations.

The six individual risk values were then summed to indicate the overall risk to each bridge. Overall or maximum cumulative risk values ranged from 0 to 13, (table 10). The maximum cumulative risk value alone, however, cannot be used to evaluate the risk for a given site. For example, several sites have a cumulative risk value of 4, which is considered "moderate", but most of that overall risk can be attributed to a high susceptibility to pier scour, which had an individual risk value of 3. Individual categories and the cumulative risk values were color coded to highlight high susceptibility to an individual risk as well as the cumulative risk (table 10). Cumulative risk values ranged from 0 to 9 for all of the sites except BN 1149, which had a value of 13. The color coding for the cumulative risk was limited to a maximum value of 9 to be more illustrative of the range of values.

Table 9.   Range of values used to assign levels of risk to bridges over tidal waterways in Alaska.

[Levels are from 0 to 3; 0 indicating no risk or not applicable and 3 indicating a high potential risk. N/A, the risk is not applicable to the category; <, less than; >, greater than]

Risk	Level			
	0	1	2	3
Storm surge	N/A	N/A	N/A	Identified risk
Tidal velocity	N/A	< ½ critical velocity	> ½ critical velocity	> critical velocity
Storage potential	none	within channel	twice bridge opening	> twice bridge opening
Channel instability[1], in feet	none	2 to 3	3 to 5	> 5
Contraction scour, in feet	N/A	> 0 to 1.5	1.5 to 3	> 3
Pier scour, in feet	N/A	> 0 to 2	2 to 4	> 4

[1]Scour risk associated with channel instability was ranked on the basis of the measured range of bed elevations determined from repeated channel surveys.

Table 10. Categories and the sum of the assigned levels of risk to bridges over tidal waterways in Alaska.

[Levels are from 0 to 3; 0 indicating no risk or not applicable and 3 indicating a high potential risk, – not evaluated. The cumulative risk values ranged from 0 to 9 for all of the sites except BN 1149, which has a value of 13. The color coding for the cumulative risk was limited to a maximum value of 9 to be more illustrative of the range of values]

Bridge No.	Bridge name	Channel instability	Upstream storage	Tidal velocities	Storm surge	Contraction scour	Pier scour	Overall risk
214	Swanson River	0	2	–	–	1	2	5
301	Klawock	0	0	–	–	–	–	0
347	Bonanza Creek	0	3	–	3	–	0	6
385	Salt Creek	0	3	2	–	1	1	7
387	Chilkoot River	1	1	–	–	1	2	5
399	King Salmon Creek	0	0	–	–	1	2	3
400	Leader Creek	0	0	–	–	0	1	1
402	Pauls Creek	0	0	–	–	1	2	3
418	Sheep Creek	0	0	–	–	1	2	3
429	Blind River	0	2	–	–	2	1	5
444	Salmon River	2	2	–	–	1	3	8
620	Ingram Creek	0	1	–	–	1	2	4
627	Placer River Overflow	2	1	–	–	1	3	7
629	Placer River Main Crossing	2	1	–	–	1	3	7
630	Portage Creek 1	2	1	–	–	1	2	6
631	Portage Creek 2	3	1	–	–	0	2	6
634	Twentymile River	2	2	3	–	0	2	9
636	Peterson Creek	0	1	–	–	0	2	3
638	Virgin Creek	0	3	1	–	2	3	9
639	Glacier Creek	1	1	–	–	0	3	5
724	Ketchikan Creek	0	1	–	–	0	–	1
732	Gold Creek	0	0	–	–	1	3	4
787	Salmon Creek Twin Lakes Drive	0	0	–	–	1	–	1
989	Sargent Creek	0	1	–	–	1	3	5
990	Russian River	2	1	–	–	1	2	6
992	Salonie Creek	0	3	–	–	2	2	7
1017	Seldovia Slough	0	3	3	–	1	2	9
1085	Hartney Bay	0	3	1	–	3	2	9
1121	Knik River NB	3	1	–	–	0	3	7
1124	Matanuska River NB	3	1	–	–	0	3	7
1127	Safety Sound	0	3	2	3	–	–	8
1149	Kenai River at Kenai Bridge	3	3	3	–	1	3	13
1188	Salmon Creek at Egan Dr	0	0	–	–	1	3	4
1197	Lemon Creek NB	0	3	1	–	2	1	7
1274	Monashka Creek	0	3	–	–	1	2	6
1385	Tununak River	0	2	–	3	–	–	5
1764	Indian Creek	0	0	–	–	–	–	0
1783	Spruce Creek	0	0	–	–	3	–	3
1863	Lemon Creek SB	0	3	1	–	2	1	7
2078	Deer Creek	0	0	–	–	–	–	0
2150	Ship Creek	1	2	2	–	1	3	9

Overall risk color code	
Low risk	0
	1
	2
	3
	4
	5
	6
	7
	8
High risk	9

High-Risk Bridges and Potential Mitigating Factors

Bridges identified in the previous section as having a high individual factor risk (value of 3) or a relatively high cumulative risk (ranking greater than 8) were investigated further to identify factors that could mitigate the risks. Several mitigating factors that were not included in the risk analyses above are discussed in the following sections for each high-risk bridge.

BN 347 Bonanza Creek

BN 347 crosses a small outlet of Safety Sound in northwestern Alaska (fig. 6) and was classified as tidally controlled. Primary risks to this bridge are the extensive upstream storage area and potential for storm surges. The upstream storage area is not a concern for normal tidal exchanges, because the maximum daily tidal fluctuation is only 2.6 ft. Storm surges, however, are common at BN 347 and can increase the volume and rate of flow through the bridge beyond what would result from the tide alone. Woody debris deposited by an undocumented storm that occurred sometime between field visits made on September 18, 2008, and September 28, 2010, was surveyed at an elevation only 2.5 ft below the low-beam elevation of the bridge. The potential for streambed scour at this bridge is mitigated by the absence of piers and by riprap armoring of the bridge abutments. The primary risks at this site are the potential for storm surge waves or ice that could impinge on the bridge.

BN 1127 Safety Sound

BN 1127 crosses the main outlet of Safety Sound in northwestern Alaska (fig. 6). Primary risks to this bridge are posed by the extensive upstream storage area and the potential for storm surges. Pier and contraction scour were not evaluated for the riverine conditions at BN 1127 because tidal exchange dominates the flow regime. Flow velocity was measured during a tidal exchange, but was not sufficient to transport sediment (table 4). The repeated cross-sectional surveys did not indicate any signs of channel instability at BN 1127 (appendix B). A small storm surge in October 2010, increased water-surface elevations at the bridge and at the Nome tide gage over two tidal cycles. Water-surface elevations increased and decreased at the same rate for the first tidal cycle, but the elevations at the bridge were higher than those at the tide gage during the second recession (fig. 8). These data are not conclusive enough to determine if the bridge constricts flow during the recession of a storm surge. Storm-surge flooding has washed over the approach road several times in the past (Mason and others, 1997), and overtopping of the road relieves the contraction of flow through BN 1127 and BN 347.

The stretch of road along Safety Sound has been armored in several locations to protect against damage due to overtopping during storms. The armoring can have a deleterious effect, however, by concentrating flow in the unprotected areas and through the bridges. Further study of the magnitude of storm surges and the associated flow velocities through the bridge is needed.

BN 385 Salt Creek

BN 385 crosses a tidal marsh on Kodiak Island that is bisected by the road approach to the bridge. The bridge and approaching roadway act as a constriction to the tidal exchanges and the primary risk identified in this study is from the upstream storage potential of incoming tidal flow. BN 385 was classified as tidally controlled. Repeated surveys of the channel cross section do not indicate channel instability at the bridge. Shear stress estimated from velocity measurements at the bridge during an outgoing tide were less than the critical shear stress required for sediment transport (table 4). Estimates of pier and contraction scour were both categorized as low risk because the minimum streambed elevation is nearly 60 ft above the bottom of the pile tips. The risk of scour associated with the contraction of the tidal flow through the bridge is mitigated by the deep pilings relative to the streambed elevations and the low flow velocities during tidal exchanges.

BN 444 Salmon River

BN 444 crosses the tidally-affected Salmon River near Gustavus, Alaska. BN 444 has a high risk of pier scour and moderate risks associated with channel instability and potential for upstream storage of incoming tidal flow. The bridge piers comprise nine individual pilings spaced approximately 3 ft apart. The pilings were treated as an individual pier in the event debris was to become lodged in the spaces between the pilings. The high risk of pier scour is the result of a flow angle of attack of 10 degrees on the piers, which was estimated from aerial photos. Pier scour was computed to equal 6.5 ft with the flow angle of attack considered, and 3.1 ft without the flow angle of attack considered. The minimum measured streambed elevation is approximately 13 ft above the shallowest pile-tip elevation. When the combined scour depths (6.5 ft of pier scour and 1.4 ft of contraction scour) were subtracted from the minimum measured streambed elevation at BN 444, the resulting minimum streambed elevation was 5.1 ft above the minimum pile-tip elevation. The estimated flow angle of attack on the piers needs to be verified at high flows to improve the accuracy of the scour computation at this site.

Bridges Crossing Rivers Draining into Turnagain Arm

The Seward Highway around Turnagain Arm crosses 12 bridges; 9 of the bridges were investigated as part of this study (fig. 10). Turnagain Arm is an approximately 30 mi long estuary that is connected to Cook Inlet and experiences some of the largest tidal fluctuations in Alaska. Streams that enter Turnagain Arm range from high-gradient streams with coarse-grained beds (sand, gravel, and cobble) to lower-gradient rivers with mostly silt beds.

The bed and bank material of the rivers at the end of Turnagain Arm is primarily silt that was deposited after the area subsided as much as 7.9 ft during the Alaska earthquake of 1964 (Bartsch-Winkler and others, 1983). In addition to local silt deposits, the rivers transport fine-grained sediment from glaciers within each river basin. Glacial-fed rivers in this area experience a seasonal cycle of channel aggradation and degradation resulting from the fluctuating discharge and sediment supply from glaciers (Conaway, 2006). Channel instability can be expected at these bridges and becomes a concern if the seasonal runoff cycle is large enough to expose pier foundations or undermine rip-rap protection at the abutments.

BN 627 Placer River Overflow and BN 629 Placer River Main Crossing

The Placer River, which drains approximately 122 mi² of glaciated basin, splits into two channels before it empties into Turnagain Arm. BN 629 crosses the current main channel and BN 627 crosses a secondary channel. Channel instability and pier scour are the greatest risks to these bridges. Both bridges were classified as tidally influenced. The distribution of flow between the channels varies, and separate hydraulic models were developed to analyze scour for each bridge. The total 1-percent AEP discharge was used in each model to analyze scour, rather than estimating the flow distribution in each channel. This approach produced conservatively high scour-depth estimates because flow would actually be conveyed in some proportion between the two bridges.

Field data were collected at these bridges on July 23, 2009. The total measured discharge at the two bridges was 9,180 ft³/s, with 4,700 ft³/s passing through BN 627 and 4,480 ft³/s through BN 629. These measurements were made after the tide had receded downstream of the bridge reaches and, therefore, do not reflect tidal or upstream storage conditions. The 1-percent AEP discharge for the Placer River is 10,150 ft³/s. Pier scour is evident in the cross sections surveyed at both bridges (figs. 11A and 11B). Measured pier scour was 5.8 ft at BN 627 and 2.0 ft at BN 629. Pier scour was measured from the concurrent ambient streambed surface (Mueller and Wagner, 2005), also called the reference streambed surface (fig. 11). The reference streambed surface was determined from data collected during the survey and

is an average of several points measured near the pier, but beyond the limits of the scour hole (Mueller and Wagner, 2005). Pier scour was computed for the July 23, 2009, measured discharge at BN 627 using the techniques outlined in this report. Computed pier scour for this discharge was 13.2 ft, more than twice the measured scour. The HEC-18 equation used in this study to compute pier scour is a design equation and as such should accurately predict scour for a given set of conditions, but errors should overpredict rather than underpredict scour (Mueller and Wagner, 2005). The discrepancy between the measured and predicted pier scour for the discharge on July 23, 2009, indicates that the predicted pier scour value of 16.5 ft for the 1-percent AEP also would be conservative (table 8).

Channel-instability and pier-scour risks are present at both bridges crossing the Placer River, but the predicted and measured pier-scour depths are greater at BN 627 because the flow angle of attack on the piers was 20 degrees at BN 627. No angle of attack was included in the pier scour computation for BN 629 because flow was observed to be aligned with the piers during the July 23, 2009, field survey. The scour analyses for the bridges are conservative because the entire Placer River discharge was modeled for each bridge. The measured discharge on July 23, 2009, was split nearly equally between the two bridges, but no information is available on the historical distribution of flow. The high angle of attack on the bridge piers at BN 627, flow distribution between the two bridges, and mobility of sediment on the Placer River will require increased monitoring because of the scour susceptibility.

BN 630 and BN 631 Bridges Crossing Portage Creek

Portage Creek flows approximately 7 mi after exiting Portage Lake and splits into two channels before entering Turnagain Arm. The main channel flows are conveyed by BN 631 and BN 630, although BN 630 crosses a secondary channel that conveys only high flows. Similar to the Placer River sites, the entire 1-percent AEP discharge for Portage Creek was modeled through each bridge. Channel instability was noted in the cross-sectional surveys at BN 630 and BN 631. Both bridges were classified as tidally influenced.

Field data were collected the two bridges on July 23 and 24, 2009. Discharges of 5,990 ft³/s were measured on July 23 at BN 631 and 990 ft³/s on July 24 at BN 630. Both measurements were made when there was no tidal influence at the bridges. The 1-percent AEP discharge for Portage Creek is 7,200 ft³/s. No pier scour was noted during these measurements (appendix B).

The minimum measured streambed elevation since 1999 at BN 630 was 11.9 ft, and the piles supporting the bridge extend to an elevation of -18 ft. The minimum measured streambed elevation since 2001 at BN 631 was 0.8 ft, and the piles supporting the bridge extend to an elevation of -22 ft.

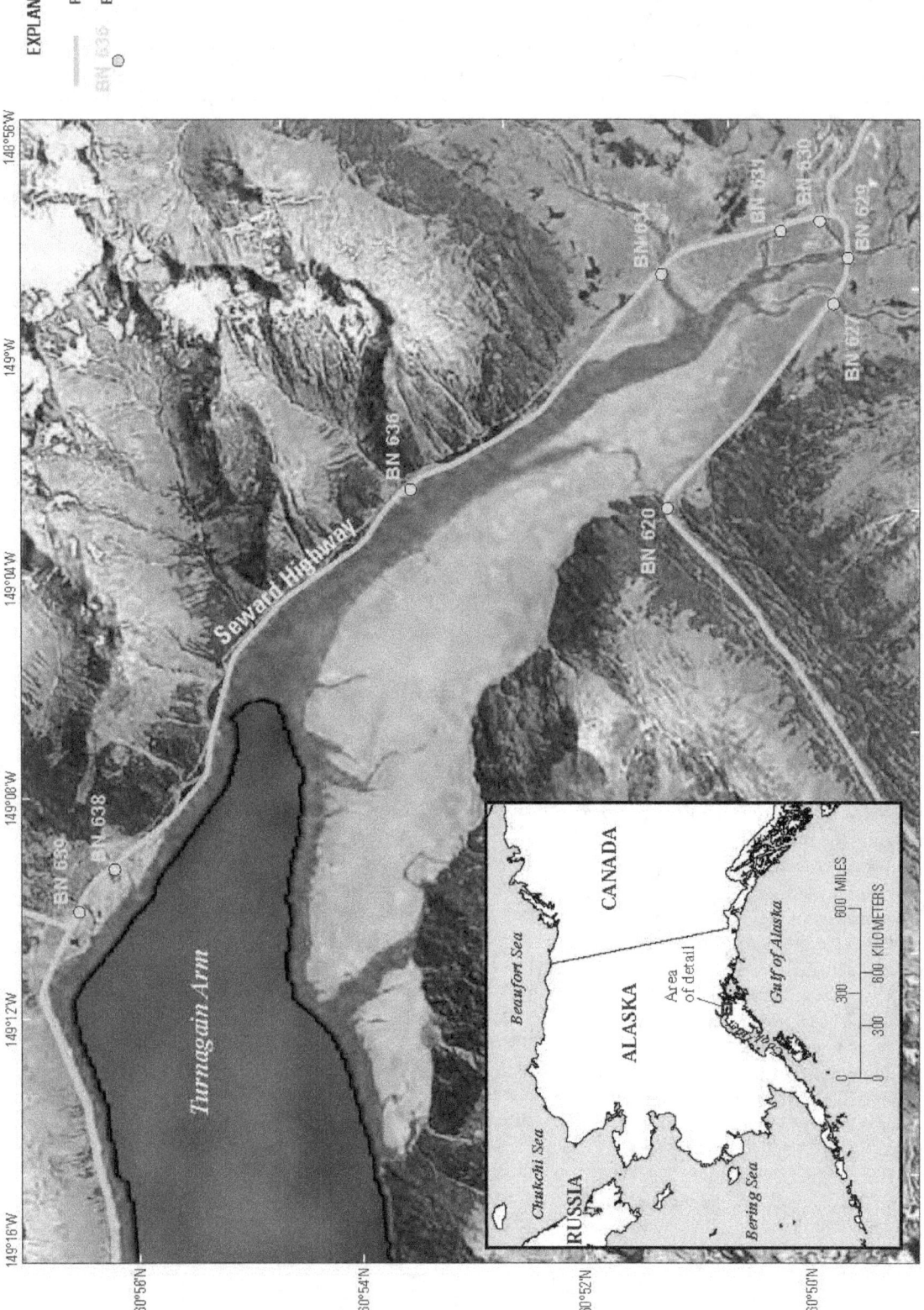

EXPLANATION

Roads

⊙ BN 638   Bridge and bridge
              number

Figure 10. Bridges over tidal waterways along Turnagain Arm, Alaska. Background image from NASA's Earth Observatory using data from Moderate Resolution Imaging Spectroradiometer (MODIS).

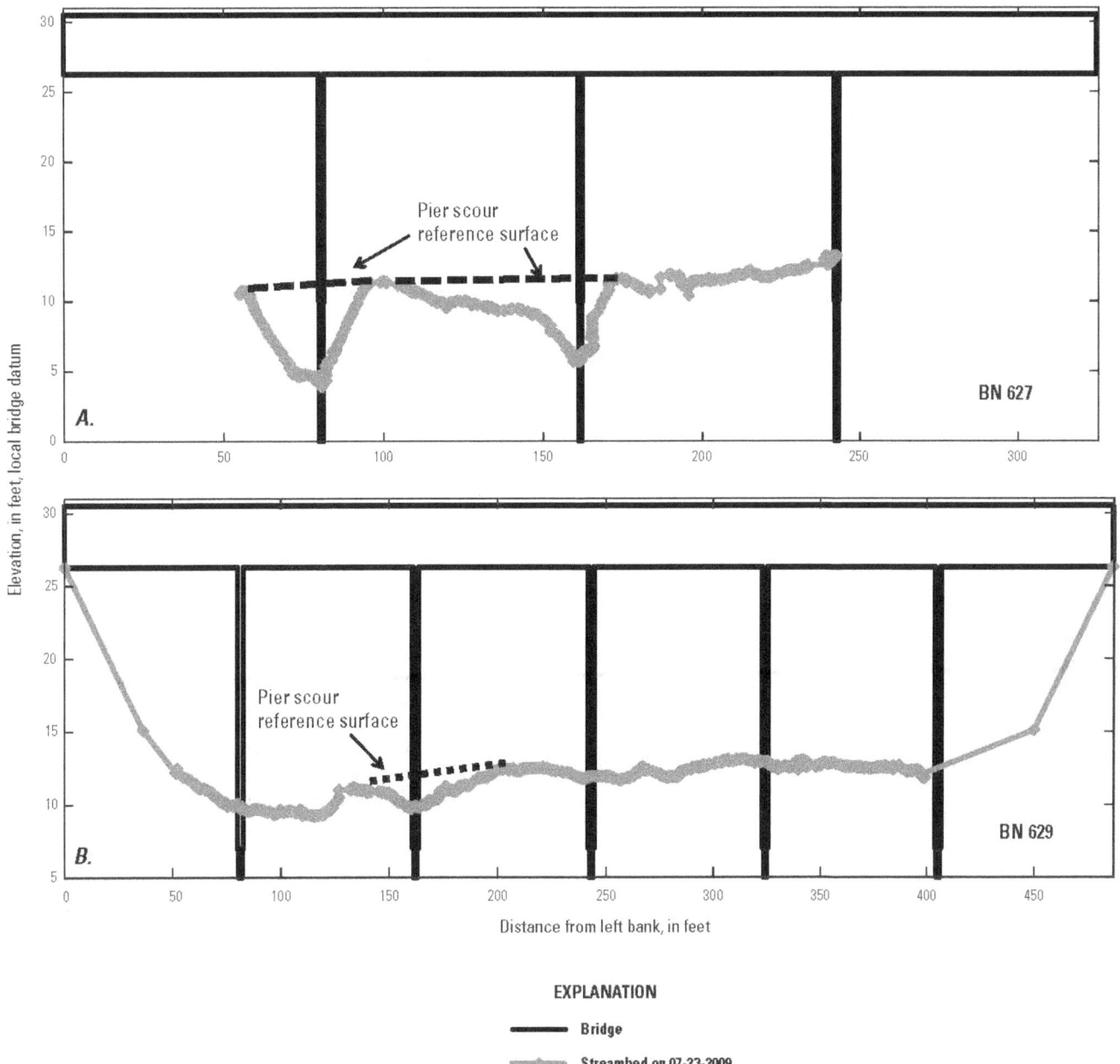

Figure 11. Surveyed cross sections at (*A*) bridge number (BN) 627 and (*B*) BN 629, along Seward Highway, Alaska. Field data were collected on July 23, 2009. Locations of bridges are shown in figure 10.

These measurements indicate sufficient embedment at both bridges. Continued biennial channel surveys of each bridge by ADOT&PF should be analyzed to determine any trend towards degradation of the streambed that could indicate the bridge is threatened by scour.

BN 634 Twentymile River

The Twentymile River, which drains a 142 mi² glacial basin, was classified as tidally influenced at BN 634. Large tidal exchanges combined with substantial storage capacity upstream of the bridge have the potential to generate high velocities during the outgoing tide. Channel instability is evident in the surveyed cross sections at the bridge. Streambed scour is concentrated along the right-bank area of the channel because the channel approaches the bridge opening at a substantial skew angle along the left-bank approach (fig. 12).

Field data were collected at BN 634 on July 20–21, 2009, to quantify the riverine and tidal conditions that might affect scour. Shear stress values estimated from velocity measurements at the bridge during an outgoing tide were greater than the critical shear stress required to initiate sediment transport (table 4). Riverine data were collected at a discharge of 14,400 ft³/s, which is approximately a 4-percent AEP flood. At this discharge, the streambed had scoured to an elevation of 1 ft at the right-bank pier. The piles supporting the bridge piers were driven to an elevation of -20 ft.

The observed scour at BN 634 and scour potential during tidal exchanges prompted ADOT&PF to increase scour monitoring at this bridge in 2011. The USGS installed a pier-mounted sonar at the pier nearest the right bank to continuously monitor streambed elevations. These data will be used in the future by ADOT&PF to evaluate the safety of the structure during high flows and periods of scour.

BN 638 Virgin Creek

Virgin Creek at BN 638 drains a small basin (4 mi²) to the south of Glacier Creek and BN 639. BN 638 was classified as tidally affected and was ranked as having a high risk of scour associated with upstream storage and pier scour. The risk of pier scour is high because the site was estimated to have a flow angle of attack of 15 degrees, which was used in the pier-scour computations. The estimated pier scour for the 1-percent AEP discharge is 4.8 ft. The minimum surveyed streambed elevation at the bridge was 6.3 ft and the elevation of the bottom of the pier is -25 ft. The pier would have 26.5 ft of embedment after accounting for the scour estimated for the 1-percent AEP discharge.

Velocity measurements were made at the bridge on September 21, 2009, during an outgoing tide after a high tide that was 3 ft higher than the MHHW. The maximum observed instantaneous velocity was 1.3 ft/s and the average velocity

during the measurement was 0.5 ft/s. Shear stress estimated from velocity measurements at the bridge during an outgoing tide were less than the critical shear stress required for sediment transport (table 4), and therefore scour during tidal exchanges is not expected.

The elevation of the bottom of the pier relative to potential pier scour depth and low velocities measured during the tidal exchange mitigate the scour risks at this bridge. In addition, cross sections surveyed at the bridge since 1999 do not indicate any channel instability.

BN 639 Glacier Creek

Glacier Creek drains approximately 46 mi² of glaciated basin and was ranked as having a high risk to pier scour because the computed scour was 6.5 ft. The bridge was classified as tidally influenced. The minimum bed elevation surveyed was approximately 27 ft above the pile-tip elevations, thus mitigating the risk associated with pier scour. Channel instability was not evident in the repeated channel soundings at BN 639, and this stability can be considered a mitigating factor pier scour that was estimated by the HEC-18 equation.

BN 1017 Seldovia Slough

BN 1017 is tidally controlled and crosses a slough that connects Seldovia Lagoon with Seldovia Bay. The bridge was ranked as high risk because of the substantial area available for upstream storage and because high velocities were measured during an outgoing tide. Shear stress values estimated from velocity measurements at the bridge during a tidal exchange were greater than the critical shear stress required to initiate sediment transport (table 4). However, both piers that support the bridge are founded on bedrock and therefore the risk of scour is low. Although the potential for streambed scour of bedrock is possible under certain hydraulic conditions (Annandale, 2007), velocities generated during tidal exchanges at this site were insufficient for this type of scour to be a concern.

BN 1085 Hartney Bay

BN 1085 crosses the entrance to Hartney Bay (at Cordova) at Orca Inlet and is tidally controlled. This bridge was ranked as high risk because of the area available for upstream storage and the estimated contraction scour for the 1-percent AEP discharge. Risks associated with the upstream storage area are low for this site because it is not subject to storm surges and the shear stress estimated from velocities measured during a tidal exchange were less than the critical shear stress to initiate sediment transport (table 4).

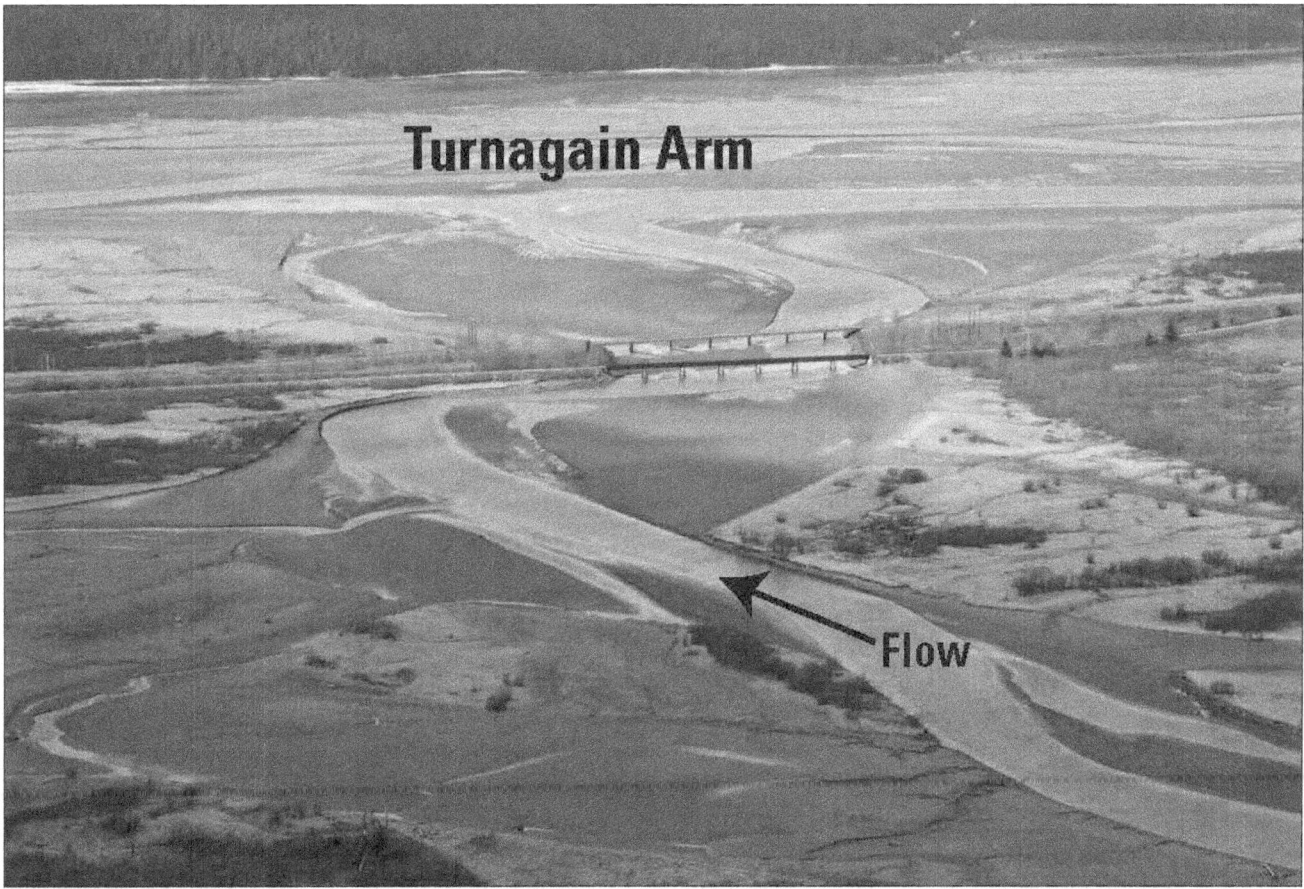

Figure 12. Oblique aerial photograph looking downstream at an Alaska Railroad bridge (upstream) and Seward Highway bridge number 634, Twentymile River, Alaska. The head of Turnagain Arm is downstream of the bridges.

The computed values for contraction and pier scour for the 1-percent AEP discharge are 4.1 ft and 3.4 ft, respectively. When the combined scour depths were subtracted from the minimum measured streambed elevation at BN 1085, the resulting minimum streambed elevation was -15.3 ft, which is only 4.7 ft above the minimum pile-tip elevation of -20 ft. Although the riverine analysis predicted both pier and contraction scour, repeat cross-section surveys at the bridge do not indicate channel instability. Given the relatively shallow pile-tip elevations at BN 1085, however, further monitoring is recommended to refine the streambed scour analysis.

BN 1121 Knik River and BN 1124 Matanuska River

The Knik and Matanuska Rivers merge into a complex interconnected system of channels upstream of Knik Arm. The Glenn Highway crosses four channels, and two of those crossings, BN 1121 and BN 1124, were included in this study. Both bridges were classified as tidally influenced. The distribution of flow through the four channels changes over time, but the channel crossed by BN 1121 currently and historically conveys the most flow (Lipscomb, 1989). This study ranked both bridges as high risk because of channel instability and estimated pier scour.

Streambed scour at BN 1121 was studied in detail by Norman (1975), who measured 6 ft of pier scour at this site on June 24, 1965, at a discharge of 73,600 ft³/s. The computed pier scour for the 1-percent AEP flow of 120,000 ft³/s equaled 9.1 ft and is considered a reasonable value based upon the value measured by Norman (1975). The 9.1 ft of computed scour would expose the upper part of the pier footings, but would not extend below the bottom of the footings. The longitudinal profile surveyed at 73,600 ft³/s indicated dunes on the streambed with amplitudes of approximately 4 ft and wave lengths of 40 ft. The channel variability noted in the repeated cross-section surveys is within the amplitude of the dunes measured by Norman (1975). No long-term degradation is evident in the repeated surveys (appendix B).

BN 1124 crosses a smaller distributary channel in the Knik-Matanuska River estuary. During a field visit on October 16, 2009, the measured discharge in this channel was less than 1-percent of the total Matanuska River discharge of 2,190 ft³/s measured at the USGS gaging station number 15284000. Gaging station 15284000 is approximately 12 river miles upstream from BN 1124. Currently the Matanuska River flows into the Knik River upstream of BN 1124 and BN 1121. Historically, however, the entire flow was through BN 1124 (Lipscomb, 1989). For this study, the entire Matanuska River discharge was modeled through BN 1124 to estimate pier and contraction scour. The computed pier-scour depth for the 1-percent AEP flow of 47,000 ft³/s was 7.2 ft, and computed contraction scour was 0.0 ft. After subtracting the computed pier-scour depth from the minimum measured streambed elevation at the bridge, the top of the pier footings would still be approximately 2 ft below the streambed. The channel instability that was noted at this site was confined to the right bank portion of the channel and represents lateral movement of the approach channel, but little change was indicated in the minimum streambed elevations. The observed distribution of flow through the bridge, measured minimum streambed elevations, and lack of vertical change in the surveyed cross sections should all be considered mitigating factors in the analysis of scour at BN 1124.

BN 1149 Kenai River

BN 1149 crosses the Kenai River approximately 5 mi upstream of the river mouth. The flow is tidally controlled and the mean tidal range at the bridge is nearly 20 ft. Velocity measurements were made on an outgoing tide on November 4, 2010, when the tidal range was 26.2 ft. Shear stress values estimated from velocity measurements at the bridge during the outgoing tide were greater than the critical shear stress required to initiate sediment transport (table 4). This was the only site in the study where the velocity measured during the tidal exchange was greater than the measured river velocity at low tide. Riverine velocities were measured at a discharge of 13,100 ft³/s on July 9, 2009. For comparison, the highest mean-monthly discharge occurs in August at a stream-gaging station located 15 river miles upstream of the bridge and is equal to 14,400 ft³/s (USGS gaging station number 15266300, Kenai River at Soldotna). The substantial temporary storage of river flow and tidal inflow during the tidal cycle results in higher measured flow velocities during the tidal exchange, compared to the flow velocities measured at low tide when flow through the bridge is only riverine.

The mobility of the streambed at this site is evident in the repeated cross section surveys made at the bridge (appendix B). The footing at the second pier from the left bank was partly exposed based on two of the soundings and nearly 6 ft of vertical change in streambed elevation was noted

between soundings at this pier. If the footing of a pier is wider than the pier and is exposed to the flow, local scour is affected by the pier's footing. For situations in which the footing is located at or below the streambed, the footing can limit local scour at the pier by disrupting the flow vortices induced by the pier (Parola and others, 1996; Melville and Coleman, 2000). When the footing is exposed to the flow, it can induce vortices that will cause scour in front of and along the side of the foundation (Parola and others, 1996; Melville and Coleman, 2000). Estimated pier scour for the 1-percent AEP discharge was equal to 8 ft, but did not account for the possible effects of the exposed footing.

The cumulative risk to scour at BN 1149 was the highest of any of the sites studied. Further study and monitoring of the streambed at this bridge is recommended to further assess the potential for streambed scour.

BN 1188 Salmon Creek

BN 1188 is on Salmon Creek in Juneau and the flow through the bridge was classified as tidally influenced. BN 1188 was ranked as having a high risk to pier scour because the estimated pier scour was 4.3 ft. The minimum streambed elevation surveyed was approximately 39 ft above each pile-tip elevation, thus mitigating the risk associated with pier scour. Channel instability was not evident in the repeated channel soundings at BN 1188. The lack of any channel instability noted in the repeated soundings can also be considered a mitigating factor to pier scour.

BN 1197 and BN 1863 Lemon Creek

BN 1197 and BN 1863 are a parallel crossing of Lemon Creek along Egan Drive in Juneau. The bridges were ranked as having a high risk due to the potential for storage of backwater-affected riverine flow upstream of the bridges during tidal exchanges. The flow at both bridges was classified as tidally affected. Velocity measurements of the outgoing tidal and backwater-affected riverine flow were made on January 29, 2010, after a high tide that was 3 ft higher than the mean high tide. Shear stress estimated from velocity measurements at the bridge during the outgoing tide were less than the critical shear stress required for sediment transport (table 4) and therefore, scour during tidal exchanges is not expected.

The risk of scour from storage of tidal and backwater-affected riverine flow at the Lemon Creek bridges is mitigated by the low measured velocities during the tidal exchange, low modeled flow velocities for the 1 percent AEP flow, and no noted channel instability. Additionally, the streambed through the bridges is armored with rip rap to inhibit scour.

BN 1274 Monashka Creek

BN 1274 crosses Monashka Creek on Kodiak Island and was classified as being tidally affected. Monashka Creek has a high potential for the temporary storage of tidal and riverine waters upstream of the bridge. A surveyed high-water mark caused either by tidal effects, riverine effects, or combination of the two corresponded to an elevation 0.5 ft greater than the modeled water-surface elevation for the 1-percent AEP discharge and was 2 ft below the low-beam elevation of the bridge (appendix A). Despite the indication of high water stage at the bridge, relatively little change in streambed elevations were noted in the repeated cross section surveys (appendix B). Pile-tip elevations were 15 ft below the lowest point in the channel, and 2.7 ft of pier scour was estimated for the 1-percent AEP discharge.

The contraction and pier scour risks at this bridge were rated as low, but the evidence of water-surface elevations near the bottom of the bridge indicate that the bridge may be undersized for the combination of flows that occur. A continuous record of water-surface elevations would help to further quantify the tidal affect at the bridge and the degree to which riverine flow is affected by backwater caused by tides.

BN 1385 Tununak River

The Tununak River bridge (BN 1385) is approximately 7 mi northwest of the village of Toksook Bay. The bridge is on a stretch of coastline that faces north into the Bering Sea and is at risk to storm surges. The 1-percent AEPWL is 11.7 ft in Toksook Bay and this elevation would be 2.8 ft below the lowest elevation of the bridge. The bridge is protected from wave action by a spit, but there is potential for upstream storage during a storm surge. No field measurements of tidal elevations at BN 1385 were made as part of this study. Further study of the magnitude of storm surges and the associated flow velocities through the bridge is needed.

BN 1783 Spruce Creek

Spruce Creek near Seward is a high-gradient stream with a coarse-grained bottom that is minimally influenced by tides where it is crossed by BN 1783. The bridge was ranked as high risk for contraction scour because 3.1 ft of scour was computed for the 1-percent AEP discharge. The minimum measured streambed elevation at the bridge was 19.9 ft. The risk associated with scour at BN 1783 is reduced because there are no piers and the abutments are constructed of sheet pile that extends approximately 20 ft below the minimum measured streambed elevation.

BN 2150 Ship Creek

Ship Creek at Anchorage is crossed by BN 2150 before draining into Cook Inlet, which has a maximum tidal range of 40.5 ft. BN 2150 is tidally controlled and has a high risk to pier scour and a high overall risk to scour based on the combined risk factors that affect the site. Velocity measurements were made on the outgoing tide on October 2, 2009, after a high tide of 11.3 ft (bridge datum). The MHHW at the bridge is 12.5 ft (bridge datum). Shear stress estimated from velocity measurements at the bridge during an outgoing tide were less than the critical shear stress required for sediment transport (table 4) and therefore, scour during tidal exchanges is not expected. . Estimated pier scour for the 1-percent AEP flow was 6.3 ft and the minimum measured streambed elevation was -6.3 ft. The threat of scour at this bridge is offset by the deep pier pilings, which extend to a depth of -135 ft.

Summary

The potential for streambed scour was evaluated at 41 bridges that cross tidal waterways in Alaska. These bridges are subject to several coastal and riverine processes that have the potential, individually or in combination, to induce streambed scour at the bridge or in the adjacent channel. These processes include, but are not limited to, storm surges, long-shore drift, tidal exchanges, sea ice, changes in sediment supply, and riverine flooding.

The degree of tidal influence at each bridge was determined to select the best approach for evaluating the potential for streambed scour. The proximity of the bridge to the ocean and water-surface elevation data collected over a tidal cycle at 33 of the bridge sites were used to identify the flow regime—whether tidal, riverine, or mixed—that has the greatest potential to induce streambed scour. The degree to which the tides affect flow through the bridge was considered to classify each site as either tidally affected, tidally influenced, or tidally controlled. Tidally-controlled bridges have a full flow reversal at every tide cycle. Tidally-affected bridges have reverse flow during some tide cycles, but tidal action is not the dominant flow condition. Tidally-influenced bridges are dominated by riverine flow and experience backwater at the bridge, but no flow reversal. By these criteria, 14 bridges were classified as tidally affected, 17 as tidally influenced, and 10 as tidally controlled. Water-surface elevations were measured through at least one tide cycle at 33 bridges and a correlation was made to the nearest tide gage. The shape of the hydrograph from the nearest tide gage was compared to the shape of the hydrograph measured at the bridge. The tidal analysis made at 33 bridges identified 12 sites where the tidal portion of the hydrograph at the bridge was asymmetric. The asymmetry is caused by the temporary backwater effect and storage of riverine flows during the incoming tide and results in a longer duration and higher discharge and higher velocities on the outgoing tide at the bridge. This scenario can, therefore, lead to scour because the potential for sediment transport through the reach on the outgoing tide is greater than that of the incoming tide.

Storm surges have the potential to affect three of the bridges studied by temporarily increasing water-surface elevations beyond the normal extent of the high tides and possibly impinging upon the bridge. The 1-percent annual exceedance probability water-level predictions for storm surge flooding were within 4 ft of the low-beam elevation at BN 347, Bonanza Creek near Nome, and BN 1385, Tununak River near Tununak. Further study and more detailed modeling would be required to evaluate the streambed scour potential of storm surges at these three bridges.

The greatest potential for scour of the streambed during the tidal exchange at a bridge is during the outgoing tide. Velocity data were collected at 10 tidally-controlled bridges during outgoing tides to determine if the velocity was sufficient to initiate sediment transport and induce scour at the bridge. The measured velocities collected during the outgoing tide were used to estimate boundary shear stress, which was then compared to the critical boundary shear stress for the sediment grain size at the study sites. Shear stress computed from velocity data collected during the outgoing tide exceeded the critical boundary shear stress required to initiate sediment transport at BN 634, Twentymile River near Portage, BN 1017, Seldovia Slough, and BN 1149, Kenai River at Kenai. The potential for scour at BN 1017 is minimal because the bridge piers are founded in bedrock. Streambed scour from the combined effects of riverine and tidal flows was observed during surveys at BN 634 and BN 1149.

Repeated surveys of channel cross sections at the bridges were compared to determine if the streambed at the site was unstable and if there was a trend in degradation or aggradation of the channel at the bridge. None of the 41 sites exhibited signs of long-term degradation or aggradation, but 14 sites were classified as having an unstable channel at the bridge.

Of the 41 bridges studied, 31 were characterized as either tidally affected or tidally influenced. The primary risk for streambed scour at these sites is from riverine flows rather than tidal fluctuations. Field data including channel cross sections, a discharge measurement, and a water-surface slope were collected at 35 bridges. The Hydrologic Engineering Center's River Analysis System (HEC-RAS) was used to calculate the hydraulic variables needed to compute estimates of scour for the 1 percent annual exceedance probability discharge. The downstream boundary condition for the models was either an energy-gradient slope or the minimum low-tide elevation. A high-flow, low-tide scenario represents the hydraulic conditions with the steepest water-surface slope and highest velocities and is therefore the most conservative estimate of scour at a tidal waterway. Contraction and pier scour were computed using recommended techniques and equations outlined in the Federal Highway Administration Hydraulic Engineering Circular No. 18 (HEC-18). Computed contraction-scour depths were greater than 2.0 ft at 5 bridges and pier-scour depths were 4.0 ft or greater at 15 bridges.

Bridges over tidal waterways are subjected to a number of processes that can threaten the stability of the structure individually or in combination. These processes include scour during tidal exchanges or storm surges, scour and erosion due to hydrodynamic loading from waves, and scour from riverine flows. The evaluated processes were ranked and summed to determine an overall risk factor for each bridge. Bridges that had high individual or overall rankings were investigated in more detail for factors that could mitigate the scour risks. Mitigating factors for sites with high risk to pier and contraction scour included deep pier depths relative to predicted scour depths, armored channels, piers founded in bedrock, and no observed channel instability in repeated channel surveys. Measured velocities during an outgoing tide that were less than those required for sediment transport was considered a mitigating factor at tidally controlled bridges where scour risk is greatest during the tidal exchange.

After accounting for mitigating factors, additional study and monitoring is needed to better quantify the streambed scour potential at nine bridges. Determination of the potential for streambed scour from storm surges at BN 347, Bonanza Creek near Nome, BN 1127, Safety Sound near Nome, and BN 1347, Kenai River at Kenai, will require more data collection and possibly hydrodynamic modeling. Validation of the flow angle of attack of the approaching flow on the bridge piers during high-flows is needed to refine the pier scour estimates at BN 444, Salmon River near Gustavus, and BN 627, Placer River Overflow near Portage. Annual channel soundings at BN 1085, Hartney Bay at Cordova, are necessary to ensure that the channel is stable and does not scour around the shallow piers. A continuous record of water-surface elevations at BN 1274, Monashka Creek near Kodiak, would help to further quantify the tidal affect at the bridge and the degree to which riverine flow is affected by backwatered caused by tides. Continuous monitoring of water-surface and streambed elevation at one or more piers at BN 634, Twentymile River near Portage, and BN 1149, Kenai River at Kenai, would improve understanding of the tidal and riverine influences on streambed scour.

References Cited

Annandale, G.W., 2007, Prediction of scour at bridge pier foundations founded on rock and other earth materials, Transportation Research Record: Journal of the Transportation Research Board, v. 1696/2000, p. 67–70.

Bartsch-Winkler, S., Ovenshine, A.T., and Kachadoorian, R., 1983, Holocene history of the estuarine area surrounding Portage, Alaska as recorded in a 93 m core: Canadian Journal of Earth Sciences, v. 20, p. 802–820.

Blier, W., Keefe, S., Shaffer, W.A., and Kim, S.C., 1997, Storm surges in the region of western Alaska: American Meteorological Society, Monthly Weather Review, v. 125, p. 3094–3108.

Brunner, G.W., 2010, HEC-RAS, River Analysis System Hydraulic Reference Manual, version 4.1, January 2010: Davis, Calif., U.S. Army Corps of Engineers Hydrologic Engineering Center (HEC), CDP-69, 411 p.

Chapman, R.S., Kim, S.C., and Mark, D.J., 2009, Storm-induced water level prediction study for the western coast of Alaska: U.S. Army Engineer Research and Development Center, Coastal and Hydraulics Laboratory (ERDC/CHL) Letter Report, 120 p.

Conaway, J.S., 2004, Summary and comparison of multiphase streambed scour analysis at selected bridge sites in Alaska: U.S. Geological Survey Scientific Investigations Report 2004–5066, 34 p.

Conaway, J.S., 2006, Comparison of long-term streambed scour monitoring data with modeled values at the Knik River, Alaska: Amsterdam, Holland, Proceedings of the International Conference on Scour and Erosion, 8 p.

Curran, J.H., Meyer, D.F., and Tasker, G.D., 2003, Estimating the magnitude and frequency of peak streamflows for ungaged sites on streams in Alaska and conterminous basins in Canada: U.S. Geological Survey Water-Resources Investigations Report 03-4188, 101 p.

Douglass, S.L., and Krolak, J., 2008, Highways in the coastal environment (2d ed.): Federal Highway Administration Hydraulic Engineering Circular No. 25, Publication No. FHWA-NHI-07-096, 246 p.

Glenn, J.S., 1994, Sensitivity analysis of bridge scour equations, in Proceedings of the 1994 National Conference on Hydraulic Engineering: Buffalo, N.Y., American Society of Civil Engineers, p. 11–15.

Heinrichs, T.A., Kennedy, B.W., Langley, D.E., and Burrows, R.L., 2001, Methodology and estimates of scour at selected bridge sites in Alaska: U.S. Geological Survey Water-Resources Investigations Report 00-4151, 44 p.

Hicks, D.M., and Mason, P.D., 1991, Roughness characteristics of New Zealand rivers: Wellington, New Zealand, Water Resources Survey, 329 p.

Julien, P.Y., 1998, Erosion and sedimentation: Cambridge University Press, 280 p.

Laursen, E.M., 1963, An analysis of relief bridge scour: Journal of Hydraulics Division, American Society of Civil Engineers, v. 89, no. HY3, p. 93–118.

Lipscomb, S.W., 1989, Flow and hydraulic characteristics of the Knik-Matanuska River estuary, Cook Inlet, Southcentral Alaska: U.S. Geological Survey Water-Resources Investigations Report 89-4064, 52 p.

Mason, O.K., Neal, W.J., and Pilkey, O.H., 1997, Living with the coast of Alaska: Durham, N.C., Duke University Press, 348 p.

Melville, B.W., and Coleman, S.E., 2000, Bridge Scour: Highlands Ranch, Colo., Water Resources Publications, 550 p.

Mueller, D.S., and Wagner, C.R., 2005, Field observations and evaluations of streambed scour at bridges: Federal Highway Administration Publication FHWA-RD-03-052, 131 p.

Murillo, J.A., 1987, The scourge of scour: Civil Engineering, American Society of Civil Engineers, v. 57, no. 7, p. 66–69.

National Climate Data Center, 2011, Storm events database: National Atmospheric and Oceanic Administration database, accessed September 28, 2012, at http://www.ncdc.noaa.gov/stormevents/.

Nelson, J.M., Bennett, J.P., and Wiele, S.M., 2003, Modeling flow, sediment transport, and bed evolution in channels, in Kondolf, M., and Piégay, H., eds., Tools in fluvial geomorphology: New York, Wiley, p. 696.

Norman, V.W., 1975, Scour at selected bridge sites in Alaska: U.S. Geological Survey Water-Resources Investigations Report 32-75, 160 p.

Parola, A.C., Mahavadi, S.K., Brown, B.M., and El Khoury, A., 1996, Effects of rectangular foundation geometry on local pier scour: Journal of Hydraulic Engineering, v. 122, no. 1, p. 35–40

Richardson, E.V., and Davis, S.R., 2001, Evaluating scour at bridges (4th ed.): Federal Highway Administration Hydraulic Engineering Circular No. 18, Publication No. FHWA NHI 01-001, 378 p.

Richardson, E.V., Harrison, L.J., Richardson, J.R., and Davis, S.R., 1993, Evaluating scour at bridges (2d ed.): Federal Highway Administration Hydraulic Engineering Circular No. 18, Publication No. FHWA-IP-90-017, 131 p.

Richardson, E.V., and Lagasse, P.F., eds., 1999, Stream stability and scour at highway bridges: Reston, Va., American Society of Civil Engineers, 1,040 p.

U.S. Department Transportation, 1988, Scour at bridges: Washington, D.C., Federal Highway Administration Technical Advisory T 5140–20, 6 p.

Wilcock, P.R., 1996, Estimating local bed shear stress from velocity observations: Water Resources Research, v. 32, no. 11, p. 3361–3366.

Wise, J.L., Comiskey, A.L., and Becker, R., 1981, Storm surge climatology and forecasting in Alaska: Arctic Environmental Information and Data Center Report, 60 p.

Zevenbergen, L.W., Lagasse, P.F., and Edge, B.L., 2004, Tidal hydrology, hydraulics and scour at bridges: Federal Highway Administration Hydraulic Engineering Circular 25, Publication No. FHWA-NHI-05-077 HEC-25, 168 p.

Glossary

Annual Exceedance Probability (AEP) discharge   Annual exceedance probability of a peak flow is the probability of that flow being equaled or exceeded in a 1-year period and is expressed as a decimal fraction less than 1.0. The recurrence interval of a peak flow is the number of years, on average, in which the specified flow is expected to be equaled or exceeded one time. Exceedance probability and recurrence interval are mathematically inverse of each other; thus, an exceedance probability of 0.01 is equivalent to a recurrence interval of 100 years.

Mean higher high water (MHHW)   The average of the higher high water height of each tidal day observed over the National Tidal Datum Epoch.

Mean high water (MHW)   The average of all the high water heights observed over the National Tidal Datum Epoch.

Mean sea level (MSL)   The average of all the high water heights observed over the National Tidal Datum Epoch. For stations with shorter series, comparison of simultaneous observations with a control tide station is made in order to derive the equivalent datum of the National Tidal Datum Epoch.

Mean low water (MLW)   The average of all the low water heights observed over the National Tidal Datum Epoch. For stations with shorter series, comparison of simultaneous observations with a control tide station is made in order to derive the equivalent datum of the National Tidal Datum Epoch.

Mean lower low water (MLLW)   The average of the lower low water height of each tidal day observed over the National Tidal Datum Epoch. For stations with shorter series, comparison of simultaneous observations with a control tide station is made in order to derive the equivalent datum of the National Tidal Datum Epoch.

National tidal datum epoch (NTDE)   The specific 19-year period adopted by the National Ocean Service as the official time segment over which tide observations are taken and reduced to obtain mean values (for example, mean lower low water, and so forth) for tidal datums. It is necessary for standardization because of periodic and apparent secular trends in sea level. The present NTDE is 1983 through 2001 and is actively considered for revision every 20–25 years. Tidal datums in certain regions with anomalous sea level changes (Alaska, Gulf of Mexico) are calculated on a modified 5-year epoch.

Appendix A.   High and Low Tide Elevations and Water-Surface Elevations at the 1-Percent Annual Exceedance Probability (AEP) Discharge at Cross Sections at 33 Bridges over Tidal Waterways in Alaska

Appendix A contains plots of surveyed channel cross sections and bridge geometry. Elevations of the maximum high and minimum low tides (since 1901), mean higher high water (MHHW), and mean lower low water (MLLW) also were plotted on cross sections at each of the 33 bridges to determine the tidal elevations relative to the bridge structure. The elevation of the modeled water surface at the bridge for the 1 percent annual exceedance probability (AEP) discharge is included for the locations where a scour analysis of riverine flow was undertaken. All plots are in the local as-built datum of the bridge.

Appendix A is available in Microsoft Excel format at http://pubs.usgs.gov/sir/2012/5245/.

Appendix B.   Multiple Cross-Section Surveys for 42 Bridges over Tidal Waterways in Alaska

Appendix B contains plots of repeated survey cross sections at the bridge for 41 sites. Most sites include data from three sources; a channel survey completed when the bridge was built (as-built), channel soundings completed by the Alaska Department of Transportation and Public Facilities (DOT), and channel soundings completed by the U.S. Geological Survey (labeled only with the date surveyed). All surveys are in the local as-built datum of the bridge.

Appendix B is available in Microsoft Excel format at http://pubs.usgs.gov/sir/2012/5245/.

www.ingramcontent.com/pod-product-compliance
Lightning Source LLC
Chambersburg PA
CBHW081402170526
45166CB00010B/3177